# HDTV:

The Politics, Policies,
and Economics of

# TOMORROW'S TELEVISION

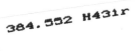
**Library of Congress Cataloging-in-Publication Data**

HDTV: the politics, policies, and economics of tomorrow's
television
  / John F. Rice, editor. -- 1st ed.
      p.  cm.
   ISBN 0-941817-08-3
   1. High definition television--United States. 2. High definition
television--Government policy--United States. I. Rice, John F.,
1955-
HE8700.74.U6H38   1990
384.55'2--dc2089-20343
                                                    CIP

First edition.

Printed in the United States of America.

Designer: Susanne Schropp.
Paintbox illustrator on cover: David O. McCloskey.
Cover illustration executed at Applied Graphics Technologies,
New York.

Published by Union Square Press, Five East Sixteenth Street, New
York, NY 10003.
Union Square Press is a division of NJSambul & Company Inc.

# HDTV:

The Politics, Policies, and Economics of

# TOMORROW'S TELEVISION

Edited by John F. Rice

Union Square Press
NEW YORK

# Contributors

**John F. Rice** is president of Rice Communications, a media consulting firm. He is former publisher and editorial director of *Videography* and *Corporate Television*, two trade publications serving the professional video industry. He was also editorial director of *Television Broadcast* magazine.

**Kenji Aoki** is managing director for NHK, Japan Broadcasting Corp.

**J.J. Barry** is international president of the International Brotherhood of Electrical Workers, which represents nearly one million employees in the electronics, manufacturing, construction, utilities, and communications industries.

The **Congressional Budget Office** furnishes the Congress with key information relating to the U. S. economy, the federal budget, and federal programs. The nonpartisan agency assists members of Congress in analyzing the interaction between the federal budget and the nation's economy, and in assessing the fiscal and budgetary cosequences of legislative actions.

**Dr. Corey Carbonara** is the new video technologies project director and assistant professor, telecommunications division, at Baylor University.

**Bob Davis** is a staff reporter for *The Wall Street Journal.*

**Dr. D. Joseph Donahue** is senior vice president, technology and business development, for Thomson Consumer Electronics.

The **Electronic Industries Association's Advanced Television Committee** includes virtually all of the major companies involved in the U. S. manufacture of studio, transmission, and viewing equipment.

**Fern Field** is a producer at Brookfield Productions in Hollywood, California. She produced *The Littlest Victims*, a CBS made-for-television movie shot in HDTV.

**Senator Al Gore**, D-Tenn., is chairman of the Subcommittee on Science, Technology and Space of the Senate Commerce Committee.

**Dr. Jeffrey Krauss** is a consultant to the cable industry.

**Barbara London** is video curator at The Museum of Modern Art in New York.

**Rep. Edward Markey**, D-Mass., is chairman of the House's Subcommittee on Telecommunications and Finance of the Committee on Energy and Commerce.

**Harry Mathias** is a cinematographer in Los Angeles and coauthor of *Electronic Cinematography*.

**Brian McKernan** is editor of *Videography*, a trade magazine for the professional video industry.

The **Office of Technology Assessment** was created in 1972 as an analytical arm of Congress. OTA's basic function is to help legislative policy makers anticipate and plan for the consequences of technological changes and to examine the many ways, expected and unexpected, in which technology affects people's lives.

**Murray Oles** is a vice president at Applied Graphics Technologies, where he is the studio manager for AGT's Manhattan SuperStudio. He has also held positions at Crosfield, Canada Inc. and Batten Graphics.

**Michel Oudin** is the vice president of engineering and development at Société Française de Production (S.F.P.) of Paris, France.

**Dr. Paul Polishuk** is president of IGI Consulting, Inc.

**Ronald Ratner** is president of Club Theatre Network, a chain of high-definition theaters and entertainment centers in Florida.

**Barry Rebo** is president and CEO of REBO High Definition Studios, Inc.

**Rep. Don Ritter**, D-Pa., is a member of the House's Subcommittee on Telecommunications and Finance of the Committee on Energy and Commerce.

**Patrick Samuel** is the chief executive officer of International HD.

**Stuart Samuels** is the vice president of Zbig Vision. He was named one of the Top 50 Producers of the Year by *Millimeter* magazine in 1988. He is the author of *Midnight Movies* (Macmillan, 1983).

**Michael J. Sherlock** is the president of operations and technical services at the National Broadcasting Company, NBC, Inc.

**Laurence J. Thorpe** is vice president of production technology, Sony Advanced Systems Co.

**Richard E. Wiley** is the chairman of the Federal Communication Commission's Advisory Committee on Advanced Television Service. He is a former chairman of the FCC, and is a partner is the law firm of Wiley, Rein & Fielding.

**Dr. Robin J. Willcourt** is president of The HD Pacific Company, founded to create medical and entertainment programming in high-definition television. Willcourt also is a practicing physician in Seattle.

# About the Cover

The graphic arts field is one area where high-definition technology is already widely used. Quantel Graphic Paintbox, a high-resolution imaging system used for print applications, utilizes a high-definition television screen as a monitor. The cover of this book was illustrated on such a system, by Paintbox artist David McCloskey at Applied Graphics Technologies in New York.

Original art can be created on the Paintbox or photographs or other images can be electronically scanned into and then modified on the system. In this case, the basic line composition was provided to Applied Graphics on a sheet of acetate, which was scanned into the machine using a flatbed scanner (the operation of which is similar to a conventional copier). The image can be seen immediately on the high-definition monitor. The image is then manipulated by using a set of functions inherent to the Quantel Graphic Paintbox. Some of the functions in the painting menu include "airbrush" and "painting," which allow the artist to work with complete control via a pressure-sensitive pen. Other functions create certain specific effects, such as mosaics or gradations. "Stencils," which are analogous to the masks that an airbrush artist uses to protect parts of an illustration while working on other areas, are created for each component of the total image that the artist wishes to manipulate without affecting the rest of the design.

The Paintbox initially provides the artist with a palette of approximately 30 colors. The artist can then create several million color combinations using varying percentages of cyan, magenta, and yellow, and can also add colors to the palette from the image itself (skin tones, for example). Colors from the palette can be mixed by eye on the screen, as well.

For this cover, David McCloskey created and combined approximately 15 different stencils to separate various components of the design. Several different functions were used to implement the design. For example, the background was created using the "gradation" function, and the glows around the grids and spheres were created by hand with partially saturated color, using the "airbrush" function. The highlights and shadings on the spheres were "painted" visually on the screen. The mosaics, which proved to be the most difficult part of the design to execute, were done with the "mosaic" function and the "crisp" function (which gave each square in the mosaic its crystallized, three-dimensional look). Then, at a high degree of magnification, the pixels were individually moved and painted until the desired effect was achieved.

The HDTV monitor allows such high screen resolution that virtually no visible degeneration occurs from the time the original images are scanned in to the time the finished output is generated. In addition, the high resolution (5,440 by 3,712 pixels) and the ability to zoom in extremely close to the image enable the artist to perform exact manipulations on minute portions of the design (such as the mosaic portions of this cover). A further advantage of the Graphic Paintbox is that, unlike other imaging systems, it works in "real time." That is, the processing time is so fast for each function that changes can be seen almost immediately on the screen.

The Paintbox stores the various versions of the image, as well as the individual stencils, in memory. Images and stencils can also be written to a magnetic tape, which is then used for archival storage. The magnetic tape can also be used on other equipment to generate a transparency or negative film of the image. For uses that do not require printer-quality output, the Graphic Paintbox also allows the image to be generated directly by a thermal printer connected to the machine.

# Table of Contents

# 1

## The Pot of Gold

*John F. Rice*

*H*DTV is pretty exciting stuff.

High-definition television, with its bold and brash promises, offers a new television system that is marked by wider screens and clearer, sharper pictures. For the average consumer waiting for HDTV to come to a store nearby, HDTV is something that will change the way he or she watches television.

HDTV, by the nature of its screen size, makes watching television more like watching a movie. The shape (aspect ratio) of HDTV sets is wider than current television sets by about a third, closer to the aspect ratio of film screens. The "high definition" is a sharper, clearer picture. More distinct than current television, HDTV is, again, closer to motion pictures than to television sets as we know them today.

Add to this the capability of CD-quality sound, and we are being promised a new kind of television that will be the next generation of home viewing. The question is "when will it get here?"

That's a good question.

Most realistic estimates indicate that HDTV will be available to American consumers in some form some- where around the middle of the 1990s. The Federal Communications Commission (FCC), which is responsible for the allocations of television channels and approvals of new technologies like HDTV, is well on its way in studying HDTV and expects to make a decision in 1993. The bad news is that there is a lot of ground to cover before an FCC decision, and even the best estimates don't guarantee that the final decision will be made within the 1993 timeframe. Nor do they guarantee that the decisions, once made, will immediately result in widespread implementation of HDTV for consumers, manufacturers, broadcasters, or any potential market.

The issues surrounding FCC testing and decision making have become just a small part of an increasingly global race to determine the future of HDTV. What was first presented as a nice new way to watch television has exploded into a morass of issues and debates that now involve companies from around the world, economic policies, and even potential government intervention.

There's more to HDTV than simply when we can watch television on bigger, brighter sets. To hear it told, the future of the U.S. economy (as well as those of other nations) hangs in the balance. HDTV development, and ultimately its availability to the world, is wrapped around a number of issues, which often appear to be totally unrelated. They include:

- development and approval of a system of broadcast transmission in the United States;
- implementation of HDTV as a production system for the creation of television shows, feature films, and other video productions;
- the future of the consumer electronics industry, in terms of which companies and which nations will take the lead (and capture the market) in manufacture and sales of HDTV television sets, VCRs, and related gear;
- the future of the computer industry, based on the fact that HDTV in any manifestation will require extensive data processing in its production, transmission, and display products, and thereby will need a heavy dose of semiconductors;
- the U.S. trade balance and economic strength (if

HDTV is developed and controlled by U.S. companies, many people believe that the U.S. could overcome its apparent weakness in the electronics field, especially in relation to Japanese strength);
* the true impact of HDTV on the communications future of the world.

Pretty heady stuff.

HDTV has become, for some, the banner issue of the future. Japan, which rightly claims its place as the originator of HDTV, has invested billions of dollars of government and private money in HDTV development. The European Community countries, while the latest to enter the foray in earnest, have also anteed up substantial investment dollars. The U.S. sees itself as the core of the future television industry, both because of its current lead in size of market and production of programming, and its feeling that the country should rightfully be the technical innovator and marketer to the world. But in America, there are only a series of unconnected and independent efforts toward technical development and market strategies.

When the dust settles, the question to be answered will simply be "who is in charge?" Many in the U.S. believe an American failure to capture and control the future of HDTV could mean economic ruin. HDTV means strength, power, employment, international trade, and yes, pride.

The Japanese are already broadcasting a limited schedule of HDTV programs every day. Plans are speeding along for delivery of HDTV signals to European homes by satellite. And announced plans for U.S. satellite systems also promise HDTV delivery of programming without waiting for broadcast-specification decisions from the FCC.

The technical decisions being made are directed at much more than just sending better pictures to home viewers. HDTV has practical uses in industry, in science, in art, in medicine, as well as in entertainment. In fact, HDTV is making strides in many of those areas already. Industrial teleconferences have been held using HDTV. Museums are using it. Medical diagnoses have been made in remote locations, and medicine has used HDTV for surgical instruction.  NASA has experimented with it. And more than a few television programs, commercials, and other productions have been completed using HDTV technology.  HDTV is both far away and here today.  Today's HDTV market is getting excited about the applications and directions of the technology as much as, if not more than, the development of it.  The pages of this book are dedicated to identifying those directions.  By defining the major issues that surround HDTV, we offer the opportunity for you to create your own analysis of the divergent areas that are at work. Here is a glimpse at the breadth of the issues, and an explanation of why HDTV development has become such a volatile debate.

By no means have we touched on every opinion or option, nor on all the specifics of the issues.  What we offer here is an overview and, hopefully, a sense that out of the divisive debate there can be a positive direction that would allow HDTV to fulfill its promise.

The debates are ongoing.  In the time that we've been putting this book together, positions have changed, new players have entered the game, some decisions have been made, some developments have been abandoned, some questions have been answered, still more have been asked.

When HDTV emerges as a defined technology, with distinct markets, full-scale applications, and winners and losers, I suspect it will be an amalgam of parts of all the proposals and positions discussed here.  This book is not a roadmap to the future, but a workbook from which many of the final answers can be derived.

And that is only half the process.

The developers and decision makers will create a system, or systems, of HDTV.  Only then will it be taken to the marketplace, where the ultimate future of HDTV will be determined.  For now, the future seems rife with potentials and good intentions.  It is only when the marketplace—the program producers, the broadcasters, the corporate users, the manufacturers, and finally the consumers—lines up to buy these products that we will see if HDTV is *the* future, or just a nice idea that never

quite fulfilled its potential.
    The potential is certainly there.

BACKGROUND AND HISTORY

# 2
# HDTV: The Video Future

*Richard E. Wiley*
*Partner,*
*Wiley, Rein & Fielding*

*M*ost Americans enjoy the motion-picture experience: a film presented on a large, wide, and eminently vivid screen. Now a new technological advance, HDTV, presents the possibility of emulating in the home the clarity and dimensions of 35mm cinematography. As such, HDTV represents the most dramatic change in the video industry since the advent of color television in the early 1950s.

But with all of its potential, this exciting advance will not appear in the marketplace until a number of complex technical, economic, and social problems have been effectively resolved. And how these issues are decided may present some very distinct risks for the United States. For example, HDTV could outmode the nation's huge investment in existing television receivers, threaten the future of its terrestrial broadcast industry, and entrench the United States as a second-class technological and economic power—especially when compared with Japan.

Each of the transmission industries that would like to deliver HDTV to the American public—broadcasting, cable, satellite, and telephone—faces serious challenges.

*Broadcasting.* The production of a larger and more visually dense television picture may require more than the six megahertz of electronic spectrum space currently allocated to each TV channel in this country. Identifying available spectrum sufficient to accommodate the nearly 1,400 TV stations now on the air could prove challenging. Still, the Federal Communications Commission in the U.S. believes that some form of advanced television service—HDTV or another system—can be introduced most quickly and economically in the United States through its existing broadcast system.

In addition, 6-MHz television sets are fixtures in almost every American home, and the FCC has concluded that no technical enhancement of the video medium should be allowed to make this investment obsolete. Thus, the commission also has ruled that once HDTV or another form of advanced television is introduced, service to existing standard TV sets must continue at least for some transition period. This is analogous to the government's ruling, 40 years ago, that color TV receivers had to be "backward compatible" with black-and-white sets.

*Cable.* The cable-TV industry transmits its programming not through the electromagnetic spectrum but, instead, by coaxial cable. Accordingly, if a broader channel is required for HDTV, cable conceivably could combine two of its channels (effectively 12 MHz) to produce an advanced image. While this might be a technical fix to the limitations of 6 MHz, it could create business problems for the cable industry. Despite the development of such new programming services as Home Box Office, ESPN, and the Discovery Channel, cable today is still largely dependent on the retransmission of broadcast signals for as much as two-thirds of its fare.

So cable probably will not move away from a broadcast-compatible regimen too quickly. Indeed, complementarity and cooperation, rather than antagonism, seem to mark the current relationship between the broadcast and cable industries with respect to advanced television development.

*Satellite.* In Europe and Japan, policy makers are planning to implement national HDTV operations transmitted by satellite directly to the home (via spectrally broader channels) rather than by terrestrially based local broadcast stations as in this country. Delivery of HDTV by direct broadcast satellite, or DBS, is certainly feasible from a technical standpoint in the United States as well. The FCC has already allocated spectrum for such a new transmission medium, and in February, 1990, a DBS joint venture of the General Motors Corp.'s Hughes Communications (a leading satellite provider), NBC, Cablevision Systems Co. (a large multisystem cable operator), and the News Corp. (a huge multimedia company headed by Rupert Murdoch) was announced. The venture, called Sky Channel, is designed to offer some 108 video channels, including both standard and HDTV signals. However, the economic viability and consumer acceptance of this venture remains to be seen.

*Telephone.* The telephone industry (and, indeed, the cable industry as well) has expressed interest in HDTV delivered by fiber optics. Now being introduced in selected areas for telephone transmission, fiber is perceived by many as an optimum transmission medium. It would replace the current copper-wire telephone plant with extremely broadband, high-capacity optical filaments. Someday, a fiber-optic system would also facilitate digital transmission of video programming (rather than the current allegedly less ideal analog mode).

But problems also exist for this medium. Fiber optics implementation would be very expensive, and some experts claim that, in particular, it may not be economically feasible to lay fiber all the way to every home. Moreover, while the telephone industry has no political barriers in building this advanced plant, it does face a number of legal and regulatory obstacles to providing video programming over fiber or existing wires. For example, both FCC rules and the 1984 Cable Policy Act prohibit a single entity from owning a telephone company and a cable system in the same market. The American Telephone & Telegraph Co. consent decree also prevents the regional Bell holding companies (which provide much of the local telephone service in this country) from offering "information services" anywhere in the United States—and the provision of video programming via fiber optics would be a prohibited information service under the decree.

Some in the telephone industry assert that if such barriers were revoked, fiber implementation would proceed more rapidly because part of the expense involved could be underwritten by programming profits. Others, especially in the cable industry, contend that fiber to the home is an unnecessary and "gold plated" solution to the provision of HDTV.

In the final analysis, I foresee a four-ring circus in which broadcast, cable, DBS, and telephone interests compete to be the predominant provider of advanced television service. In the meantime, efforts continue at the FCC to select a new terrestrial broadcast standard that also might be compatible with video delivery by alternative media (such as cable).

In 1987, the FCC established an Advisory Committee on Advanced Television Service (which I chair) to look into the technical, economic, and spectrum trade-offs involved in establishing a new transmission standard. In this regard, the committee has also been examining whether additional spectrum (beyond 6 MHz) could be made available to broadcasters if needed to deliver HDTV. Such added capacity might come from lessening the interference protection criteria and mileage separations between UHF stations mandated under the nation's current TV standard.

After two years of planning, the Advisory Committee is about to direct the laboratory testing of various advanced television system concepts (including HDTV) that have been introduced to date. Initially, 14 proponents stepped forward with some 23 different concepts. Through mergers and attrition, only six proponents and eight proposals remain as of May, 1990.

Initially, the proposals fell into three categories:

- Systems that offer *enhancement* to today's standard TV (e.g., some improvement in resolution, a wider aspect ratio, etc.), but that do not require additional frequencies;
- 9- or 12-MHz systems that *augment* the current picture with HDTV details; and

- 12-MHz *simulcast* systems that broadcast standard television on one 6-MHz channel (the theory being that HDTV, in fact, could be accomplished in 6 MHz if engineers were permitted to start with a "clean" channel, without the technical artifacts contained in the current standard).

In March 1990, the FCC directed the Advisory Committee to discontinue its consideration of augmentation systems and, with respect to HDTV, to focus exclusively on the simulcast approach. However, the commission also ordered the committee to continue to examine enhanced system proponents. While FCC Chairman Alfred Sikes has expressed a preference for simulcast HDTV, he also made clear that technical, cost, and consumer acceptance factors may dictate the ultimate selection of an enhanced system standard.

Based on the Advisory Committee's recommendations and other public and expert opinion, the FCC intends to make its standards determination in mid-1993. This would permit advanced receivers to be sold to the American public as early as 1994 or 1995. The agency is also likely to establish a technical "interface" between its broadcast standard and cable-TV delivery.

A final, extremely important issue relating to the introduction of HDTV is what, if anything, the government should do to promote U.S. industry involvement in this new technology. Because our country undoubtedly will be the largest consumer of advanced television sets and programs, various policy makers—including FCC Chairman Sikes and Rep. Edward Markey (D-Mass.), chairman of the House Energy and Commerce Subcommittee on Telecommunications and Finance—argue that the U.S. should be a significant factor in the program production and industrial aspects of HDTV as well.

Without doubt, the United States is destined to be the world's leader in HDTV software. But the nation unfortunately has slipped in recent years in technology and manufacturing. Our consumer electronics industry is virtually nonexistent, and only one American-owned manufacturer of television sets, Zenith Electronics, remains. Many policy makers see HDTV as an opportunity to reverse this decline not only in the video field but in other high-technology areas as well. While the potential of HDTV is not assured, it could be an economically important business someday. It also may have significant interrelationships to other industries, such as computers (where the United States has about 70 percent of the world's market, but is increasingly vulnerable to foreign competition), semiconductors (where the U. S. has lost its leadership position, particularly in the crucial memory chip, or DRAM, market), and display technology (which may be crucial to future advanced commercial and defense products).

For this reason, some advocates have recommended that the government provide active support to domestic companies attempting to develop HDTV and other high-technology services. Aid might take various forms: standard setting, which, as indicated, is likely to happen with respect to HDTV; changes in or interpretations of antitrust law that would allow U.S. companies to combine research efforts into expensive products with long lead times; changes in tax law to promote research and development; and funding to jump-start dormant private-sector initiatives.

Other voices have suggested that these efforts, especially government investment, would not be effective and, in any case, should not be limited to a single new technology such as HDTV. Moreover, since the federal budget is so restricted, funding might be hard to come by even if the philosophical hurdles were surmounted. Perhaps only the Defense Department's Advance Research Products Agency has had the ability to provide support to domestic firms attempting to develop advanced display technologies, such as a flat television screen that could be hung on the wall like a painting (and that would have many other scientific, medical, and defense applications), and new microprocessor innovations relative to receiver design and development.

Whatever happens with regard to HDTV, new government policies may be needed to encourage increased R&D in the United States, strengthen the links between basic research and manufactured goods, lower the cost of capital to a level similar to that available to foreign corporations, and, perhaps most of all, improve the national system of education so that American workers can truly compete with their counterparts abroad. HDTV might well provide an important opportunity in all these areas, but it is only one new technology to which such policies could be applied.

Myriad challenges must be confronted before it can be determined where the U.S. (and other nations as well) are headed in the advanced television field. Fortunately, federal agencies and a host of private-sector entities are today placing substantial emphasis on the resolution of these problems. All this activity and attention ultimately may bring our citizens not only clearer television but also a healthier and more vigorous high-technology economy.

Copyright American Lawyer Media, L. P. This atricle is adapted from the author's "The Video Future: Advanced TV Not Just For Couch Poatoes," in *Legal Times*, May 14, 1990.

# 3

# High-Definition Television:
# A Major Stake for Europe

*Patrick Samuel*
*Chief Executive,*
*International HD*

*T*he advent of high-definition television constitutes a major stake for Europe, to which the
continent has made a major commitment and
developed a coherent and reasonable approach, which
we feel is acceptable to all the concerned international
partners.

The stake is triple: It is simultaneously economic,
strategic, and cultural.

As for the economic stake, the arrival of HDTV will
require the renewal of the worldwide stock of consumer
display and recording equipment.

In this respect, the figures speak for themselves: The
world stock of television sets in service today verges on
800 million units and according to certain American
studies, which may be somewhat optimistic, the market
for high-definition receivers and videocassette recorders
should amount to up to $330 billion in the period from
1990 to 2010. The regions with the greatest purchasing
power represent about 80 percent of the world consumer electronics market, of which almost 30 percent is
attributable to Western Europe.

Now, while it is true that the market is dominated by
Japanese manufacturers, European manufacturers are
also a major part of this industry: first, because it contains

two companies of global size (Philips and Thomson Consumer Electronics); and, secondly, because it is largely dominant in its own television-set market (about 80 percent of the sets sold in Europe are manufactured by European companies).

Although its volume is much more limited, the worldwide market of professional audiovisual equipment is not any less important for Europe. If Europe were absent from this sector, those who attempted to establish a monopoly in it would control the programming industry worldwide, which would not be desirable for anyone. They would also impose their distribution standard on those countries who have not defined one. Europe, however, believes that the development of a distribution standard cannot be separate from that of a corresponding production standard. While it would be absurd to deny the current strength of Japanese industry in this sector, there are also enterprises in Europe (TVE, BTS, Rank Cintel, etc.) whose medium size does not preclude a high degree of know-how and a very dynamic quality. These enterprises have developed a close cooperation that has already allowed them to develop the majority of necessary high-definition filming, shooting, recording, and processing equipment.

The third facet of the economic stake connected to the battle of HDTV is the electronic components sector, which includes optical components, tubes for cameras and receivers, and above all, semiconductors. It must be remembered that as the content of semiconductors in current television receivers has risen to 30 percent, it will rise to 70 percent for improved television, and to 90 percent for high-definition receivers. European production of semiconductors amounts to about 10 percent of global production. Philips and the SGS Thomson Group, who both belong among the frontrunners of the 15 premier global producers, decided in 1988 to join their efforts with those of the Siemens Group within the framework of a program called Jessi (Joint European Submicron Silicon), whose objective is to allow Europe to attain a technological level comparable to that of the United States and Japan within the next 10 years. They know that the manufacture of color televisions offers the

European component industry a massive opening, which would generate financial resources to feed the research and development effort needed for professional and defense electronics. In other words, the maintenance of a healthy and strong consumer electronics industry is, for Europe, a condition for maintaining its strategic freedom in an essential area: microelectronics.

Europe's strategic interest arises naturally from the preceding observations. In effect, the semiconductor industry, whose future is tied to that of HDTV, is a compelling part of a country's defense and of its independence, since it supplies the means to have modern systems of arms, aeronautics, and high-performance space equipment. It is necessary to add high-definition display technologies, which naturally can be applied to the sectors of radar screens, monitors, simulators, and other highly technical equipment. To be open to the world, to be a dependable partner in the eyes of its allies, Europe needs to be confident of its own capacity for independence. In this sense, for Europe, the battle of HDTV also signifies the refusal of all forms of technological vassalage.

For Europe, the cultural interest attached to the battle of HDTV is not separable from the technological stake.

A technological collapse would lead to a diminution of its scientific knowledge, and of its know-how in sectors where artistic creation is involved. Just as serious, technological vassalage would be accompanied by a disappearance of its programming industries, which, since today's television is an important medium of cultural transmission, sooner or later would lead to a true cultural vassalage. Of course, Europeans are perfectly conscious that the fact of having developed its own technology as far as HDTV is concerned will not protect it in every instance from such an eventuality. But, on the other hand, the effects of an eventual technological defeat are certain. Technological control in the domain of HDTV will not exempt Europeans from endowing themselves with a strong and structured programming industry.

Precisely, the arrival of HDTV offers Europe a

chance to seize control of its technological future. On one hand, because the arrival of HDTV will be accompanied by technical improvements in image as well as in sound, it will constitute a call for an increase in programming quality. In the very competitive context that exists today in Europe, quality could be a medium "to make a difference" to gain in influence. HD also will allow the introduction of the artistic demands of film production to the practice of video. Therefore, an HD audiovisual style will exist, whereby the intrinsic level of quality will constitute a stimulant for creators, for producers, for broadcasters, and of course for the public.

HD also offers Europe the opportunity to give itself a common audiovisual space. Today, European televised productions are not economically competitive because they are not able to recoup their costs, since the European audiovisual market suffers from linguistic barriers and a heterogeneity of technical standards. The new official MAC (Multiplexed Analog Components) standards are common to all of Europe; they allow us to surmount the linguistic barriers through the multiplicity of sound channels. Eventually they will offer Europe the possibility of selling less expensive productions that are already profitable in its immense market of 350 million television viewers.

Europe's massive commitment is first political. The importance of the stake, at the same time economic, strategic, and cultural, which constitutes the advent of HDTV, has not escaped the EEC authorities and the higher leaders of the member states. Some of them consider HDTV in the same way as the conquest of space, as one of the major technological challenges of the last quarter of this century.

Launched in 1986, at the initiative of French President François Mitterrand, the EUREKA 95 program allowed the appropriation of financial resources, both national and EEC, public and private, needed for a research and development effort unprecedented in volume and intensity.

From 1986 to 1989, the EUREKA 95 industrial consortium, which consisted of the participants listed in Table 3-1, succeeded, at the price of a $350 million investment, to develop, in record time, all the component equipment in the line of high-definition images and sound. The demonstrations carried out at Brighton, England, in October, 1988, and in Berlin in August, 1989, made the rest of the world take notice of the strength of the European effort.

**TABLE 3-1**
**EUREKA 95: List of Participants**

| | |
|---|---|
| Belgium: | Barco Industries |
| France: | Angenieux |
| | CCETT (TDF and France Telecom) |
| | S.F.P. |
| | Thomson S.A. |
| Finland: | Nokia |
| West Germany: | B.T.S. |
| | F.T.Z. (DBP Research Center) |
| | Fuba |
| | Grundig AG |
| | Heimann GmbH |
| | Heinrich-Hertz Institut |
| | Intermetall |
| | P.K.I. |
| | Rhode and Schwartz |
| | Studio Hamburg |
| | Schneider |
| | Universität Dortmund |
| Italy: | R.A.I. |
| | Seleco |
| Holland: | N.V. Philips Gloeilampenfabrieken |
| | N.O.B. |
| Sweden: | Swedish Telecom Radio |
| United Kingdom: | A.V.S. |
| | B.B.C. |
| | British Telecom |
| | I.B.A. |
| | I.T.V. Association |
| | Quantel |
| | Rank Cintel |

In 1990, the second phase of the project began. It will be linked to an effort of $500 million over 3 years, and will grant priority to perfecting integrated circuits, to the digitization of recording equipment, to finalizing HD tape-to-film transfer equipment, and the achievement of quality professional monitors. Portable cameras, post-production equipment, projection systems of all types, and widescreen displays, will also be the object of important efforts.

The European EUREKA 95 will be completed by a Franco-Dutch program in which Philips and Thomson will share in a joint effort amounting to over $3 billion over five years.

This development is not confined to Western Europe. President Gorbachev has repeatedly stated that it is necessary that the standard adopted by Europe be the same one that is adopted by the communist world. And the USSR, since the beginning of the HDTV battle in 1986, has indicated support for the principles of evolution and compatibility that govern the European approach.

If Europe vigorously opposed the Japanese proposal, advanced in 1986 at the International Radio Consultative Committee (CCIR) convention in Dubrovnik, Yugoslavia, it is not from coldness, nor from a spirit of denigration, nor from an effort of isolation. Europe is conscious that the Japanese proposition rests on a very advanced technology, but it considers that it would lead to a brutal technological break for Europe, unacceptable for producers and broadcasters, as well as for consumers.

It also believes, as stated earlier, that it is not possible to separately treat the problems of production and distribution, since it is true that programs, no matter how attractive they are, must first be distributed and seen by the greatest possible number of viewers, at the most economically acceptable conditions for them.

*In the domain of production,* Europe has perfected a hierarchical family of evolutionary standards, of which the most elaborated version is a candidate to someday replace the 35mm film medium as the unique world standard. The European proposition is composed

therefore of three successive levels of increasing quality
standards:

- the first, HDI (Interlaced High Definition), uses
  interlaced scanning and analog processing;
- the second, HDQ (Quincunx High Definition), uses
  progressive scanning and digital processing, but the
  information output is reduced compared to an interlaced image;
- the third, and final, consists of a progressively
  scanned image, but this time without degradation.

The choice depends on several concerns:

First, the Europeans consider that in the long term
only progressive scanning can claim to equal the quality
of the 35mm film standard, the only true worldwide
high-definition standard today.

Second, they consider that the future world high-
definition standard must be compatible with both 35mm
film and current production standards, particularly the
4:2:2 digital video standard adopted long ago by the
international community.

Certainly, incompatibility does not prevent convertibility, but the latter requires complex processes,
because of interpolations made necessary by the absence of frequency synchronization. And conversions
are expensive because of the higher and higher costs of
equipment that becomes more sophisticated year by
year. The Europeans are perfectly conscious that the day
where images filmed by a progressively scanned camera
can be transmitted without degradation is still far away,
and that is why they have chosen to place importance on
compatibility and convertibility along with other standards.

*In the domain of distribution*, the European approach is also characterized by realism.

First, Europe has developed a unique standard of
satellite distribution, the D-D2-MAC (Multiplexed Analog Components), true in its very conception to the CCIR
4:2:2 digital production standard. D-D2-MAC allows the
distribution of an image that is as full as is theoretically
possible today by both terrestrial stations and cable. It

also can distribute high-quality stereophonic sound and complex data (subtitles, teletext, etc.). These advantages are principally the result of one of the characteristics of the MAC standard, the separate digitization of the luminance (brightness detail) and chrominance (color information) components.

Secondly, the Europeans have perfected a distribution standard in high definition, the HD-MAC, directly compatible, as its name indicates, with today's D-D2-MAC, which allows a very subtle and precise reproduction of moving images. Thanks to its compatibility, old television receivers will be able to pick up programs transmitted in accordance with the new standards; in other words, transmissions on HD-MAC will be able to be received directly by current MAC receivers. Europe believes that compatibility is a commercial necessity that cannot be ignored.

Considering the rising costs of necessary equipment, it would be futile to try to interest a producer or broadcaster to implement new technologies if the target market of those who would buy HD receivers in the early years of service amounts only to a small number of viewers.

Conversely, viewers will not buy HD receivers if there is no attractive programming available. The European approach is meant to be realistic, in setting apart the short term and the long term and in permitting us, thanks to the concepts of progressivity and compatibility:

- to not wait for the perfection of display devices specifically for HD to improve television;
- to use current standards as a base of departure from which to move toward HDTV;
- to decide upon, as the ultimate objective, a global production standard truly worthy of replacing the 35mm film standard, while giving priority in the immediate future to defining a standard to be shared by countries using 50Hz and 59.94Hz frequencies;
- for producers, to begin immediately to write off HD productions by showing them to a larger audience than would be possible if they were the only

possessors of high-definition display equipment;
- to carry out, from now on, an important technological jump, a preferable solution in terms of viewing than a patching up of traditional standards, which would be followed by an unacceptable technological rupture.

By refusing in 1986 to allow the imposition of a true technological "diktat" on the international community, Europe allowed the debate to be put back on its healthier and much more favorable foundations, to a true cooperation between nations and regions. Certainly, everyone agrees today to recognize the diversity of technological and economic contexts, which will not permit, at least until the middle term, a unified HDTV distribution standard on a global level.

On the other hand, worldwide unification of standards is still a major objective for Europe in the image-production domain.

Certainly, Europe stays open to all negotiations, but always remains resolved not to sacrifice its own legitimate interests. In particular, research alliances, notably with the Americans, have always been one of the characteristics of the European approach. The accord signed in 1990 between NBC, David Sarnoff Research Center, Thomson Consumer Electronics, and Philips North America, is a significant example of this spirit.

If it is true that the Europeans, without wanting to close themselves to the American culture, are waiting for more reciprocity on the American side toward European programs; and if it is true that the Europeans intend to preserve their cultural identity, there still exists between them and the Americans a common technological philosophy based on the concept of compatibility, and also strong complementary features (Europe can benefit from the advance of the U.S. semiconductor sector, and the U.S. can benefit in return by the research results conducted in the framework of the EUREKA program). It is simply in their interest to act together to stop a single country, no matter how talented, from establishing a veritable monopoly in the electronic activities sector.

It is difficult to say who will win the battle of HDTV,

but the Europeans have already bitterly denied the famous interpellation of K. Matsushita, chairman of the Japanese electronics giant Matsushita Electronics (Panasonic), according to whom "they [the Europeans] have already lost the struggle because they carry in their minds the very cause of their defeat." Europe does not claim to be a fortress. It simply claims to be itself. It is open, but it is not offering itself to burglars.

# 4

# HDTV In Japan:
# Present and Future Prospects

*Kenji Aoki*
*Managing Director,*
*NHK, Japan Broadcasting Corporation*

*T*he number of television sets in Japan at present totals an estimated 70 million units, and the figure climbs to some 140 million for the United States. About 200 million units are in use in China (although the majority of these are of the black-and-white variety). On a worldwide scale, the number of TV receivers in current use is believed to total approximately 700 million units.

These figures indicate the potential for a gigantic new industry. If all of the 700 million TV sets were replaced today by HDTV sets, the economic impact would be enormous—and beyond the calculation skills of industry analysts. The Japanese Ministry of Posts and Telecommunications forecasts the emergence of a $100 billion industry in Japan by the next century, based on the assumptions that consumers will convert their present TV sets to HDTVs and that spin-off industries will evolve, such as a new HDTV software industry. This is a prediction for Japan alone.

Looking at it from a worldwide perspective, a complete conversion to HDTV would give rise to a mammoth industry worth well over $1 trillion. This would be great news not only to consumer electronics manufacturers but to the world economy as a whole, which could expect revolutionary changes as a result of the birth of a major new industry. This is the main issue underlying the current HDTV debate heating up around the world. Japan, the United States, and Europe are paying close attention to this debate because of a key question: Who will take the initiative to handle world demand for a $1 trillion industry?

While it is important to discuss the HDTV issue from a macroeconomic point of view, it is also essential at this time to carry out a thorough discussion of the potential of HDTV as a medium. HDTV boasts picture quality comparable to that of 35mm film, its sound quality is CD-comparable, its pictures are easily convertible to those of other media, and, above all, it has diverse application potentials for transmission and record keeping. All these attributes point to the fact that this technology can occupy a top place as a medium in an increasingly information-oriented society. HDTV represents not merely an advanced television format for the next generation of broadcasting; it opens up the door to a wide range of applications in other industries as well.

After naming its HDTV system "Hi-Vision" (which has 1,125 scanning lines and a 60Hz field rate), NHK initiated one-hour experimental broadcasts on a daily basis in June 1989. Since Hi-Vision programs are transmitted via satellite in the Japanese MUSE direct broadcast satellite system, the spread of Hi-Vision depends largely on that of direct satellite broadcasting.

NHK operated its direct broadcast satellite (DBS) in the NTSC (current television in the U.S. and Japan) format on an experimental basis from 1987 until June, 1989, when it then began full-scale broadcasts on two channels, charging each household 930 yen (approximately $7) as a monthly viewing fee. In the first fiscal year, NHK's DBS service operated in the red, losing some 20 billion yen (about $140 million). It is expected to take seven years before NHK can operate at the breakeven level. Regarding satellite dish ownership, NHK estimates that the number will grow to 2.3 million households by the end of the 1989 fiscal year, 3.3 million by 1990, and 5.7 million by the end of 1992. The number of satellite subscribers is most likely to exceed 10 million households by the end of the 1995 fiscal year.

The BS-2b, a broadcast satellite used by NHK, will be succeeded by a next-generation satellite, the BS-3a, in 1990. The new satellite is equipped with three transponders with 120-watt power output, two of which are to be used by NHK. The remaining transponder will be allocated to the Japan Satellite Broadcasting Co., which will start its own satellite broadcast business. NHK plans to augment its current service by offering additional caption, database, and facsimile broadcasting and PCM (digital) broadcast services. However, NHK's ultimate goal in satellite broadcasting is Hi-Vision, which automatically makes the wide spread of DBS an essential basis for Hi-Vision broadcasting.

As previously mentioned, NHK provides NTSC satellite broadcast service through two channels: Channel 1 exclusively covers world news and sports, and Channel 2 offers cultural and entertainment programs. This genre-per-channel concept is part of an effort by NHK to design a 24-hour-long programming format that is distinguishable from "terrestrial programs."

In June 1989, a daily one-hour Hi-Vision broadcast got underway on an experimental basis. In addition to live sports (which are the primary components of HDTV software packages), original HDTV programs produced both in Japan and overseas are also shown. These programs are also displayed to the public at Hi-Vision exhibits set up at about 100 sites around the country, including train stations, department stores, post offices, broadcast stations, manufacturers' commercial exhibit sites, and city halls. A full-scale Hi-Vision broadcast service, to be offered jointly by NHK and commercial broadcasters, is scheduled for the fall of 1991. A spare channel of the BS-3b, a back-up satellite for BS-3a, will be secured exclusively for Hi-Vision broadcasting.

There have been positive signs from the general public. In September 1989, NHK broadcast the Seoul Summer Olympic Games in the HDTV format for a total of 73 hours. Hi-Vision receivers were set up at 81 public exhibit sites all over Japan, attracting a total of 3.72 million people to the world's first Hi-Vision broadcast. A survey conducted at this time revealed that 93 percent of the people watching the Hi-Vision broadcast were interested in Hi-Vision, and 75 percent replied that they wanted to buy Hi-Vision receivers. Forty-eight percent of those polled said they would buy a Hi-Vision set if the price was within the range of two to three times the retail price of current TV sets.

The most important item to the consumer is the receiver. A 36-inch home HDTV receiver is likely to be on the market as early as the fall of 1990 for a price of about 1.5 million yen ($10,500). The price tag, however, should come down to about 500,000 yen ($3,500) within the following five years. Based on hypothetical prices, the following sales forecasts for HDTV receivers have been drawn up: By the end of 1992, the year of the Barcelona Summer Olympic Games, 300,000 units are expected to be in use. By 1994, the figure will surpass one million, and then reach the five million level in 1997. By the beginning of the 21st century, there will be 10 million Hi-Vision sets in Japan, according to current projections.

But this forecast is not without preconditions. There are still many types of HDTV equipment waiting to be developed beyond the prototype stage and major efforts are underway to meet this challenge. For example, equipment that down-converts video images from HDTV to NTSC is close to being completed.

Once commercialized, a home down-converter will retail in a range between $100 and $150. A home HDTV VCR is targeted for marketing by the end of 1991. Developing a flat-panel display and upgrading a system for converting film to or from video are just some of the many other tasks that still need to be addressed.

The impact of HDTV will be substantial, resulting in high-quality video pictures and a wide filmlike screen. But it will also result in a considerable diversification of applications, creating a new video/image system for the 21st century. It is expected that HDTV will have a substantial influence on people's lives, culture, and business as a whole.

Naturally, the production of software will not be limited solely to TV programs. In addition to simply broadcasting HDTV productions on TV, a series of activities that would never have been possible in the realm of the conventional TV have become realistic—all because of the fine quality of the HDTV picture. Examples include organizing an exposition or trade show utilizing HDTV software, or publishing a book using high-definition pictures converted to 35mm. The present situation regarding software production is evidence that broadcasting is only one of the many applications of HDTV.

A major focus of NHK at present is the organization of large international coproductions in HDTV. "Ginger Tree," which NHK coproduced with BBC and WGBH in 1989, is a typical example. After its completion, the program was converted to PAL and broadcast in Britain first. It will be later shown in both NTSC and HDTV in Japan. American viewers will also be able to see the program in NTSC.

Another joint work already completed is the production of a record-setting 15-hour-long HDTV video of *The Nibelung Ring*, an opera in four parts written by Wagner, which was performed in November, 1989. In collaboration with the Bayer National Opera Theater of West Germany, NHK recorded all four opera performances in HDTV, to be broadcast later in Japan. The plan under current discussion calls for screening the opera video in an HDTV theater where a limited number of viewers will be able to savor the art of opera in an elegant and relaxed setting.

Another possibility is the sale of HDTV disc recordings of the opera series and/or videocassettes converted into NTSC. Such artistic recordings in HDTV will prove to be priceless assets in the future.

NHK broadcast about 400 hours of HDTV programs in 1989. In 1990, 600 hours of programming are planned. And the number will continue to grow, from 2,000 hours in 1992 to 4,300 hours in 1998. Opinions differ regarding the type of programs most appropriate for HDTV broadcasts, but the primary ones will be sports such as baseball and football. In my opinion, however, HDTV will also enhance news broadcasting with its full scope of features. NHK is currently studying how best to apply the high-definition image to a news broadcast.

Conventionally, two types of cameras, NTSC and HDTV, are used for sports event coverage when broadcasting in both NTSC and HDTV; recently however, only HDTV cameras have been used for shooting, with the video pictures simultaneously down-converted for broadcasting in NTSC. A plan is gradually being implemented at NHK to replace all the NTSC cameras with HDTV cameras in large studios, in order to tape entertainment programs in HDTV and down-convert for conventional broadcasts.

In Japan, HDTV has also been applied to many industrial areas. Medicine is one area that is attracting a high level of attention. A microscopic brain surgery operation was recorded with an HDTV camera for the first time at the Shinshu University Hospital in December, 1987. By plugging an HDTV camera into a microscope and projecting finely detailed pictures of the surgical process on a large screen, it was possible for a number of the hospital's doctors and medical students to view the surgery, which would have otherwise been available only to the physicians involved.

Furthermore, another brain surgery operation was videotaped at the same hospital in February, 1989, but it was taken in 3D this time. Linking two HDTV cameras to both the microscope's right and left eye pieces, pictures were then recorded simultaneously. The educational effect far exceeded expectations; witnessing the wonder of a human body that is offered through a vivid high-definition picture of a brain was an emotional experience.

The application of HDTV in the world of art is also making progress. The Gifu Art Museum in Gifu Prefecture recently celebrated the opening of its HDTV Art Gallery in April, 1989. The HDTV system there is designed to allow guests and professional researchers free access to all of the art pieces in the museum's collection; still pictures of all the museum's paintings were taken using an HDTV camera and then stored in a CD-ROM database. In addition to reading written descriptions and hearing oral commentaries on the paintings, viewers can also magnify any portion of the picture, which is projected on a large screen. None of these functions had ever been possible before at any conventional museum.

As a result of the introduction of the new system, art lovers and other interested people have swarmed to the art museum. Attendance hit a record in the first month of the operation of the HDTV Art Gallery. The number of visitors in that month reached the same level as that of the entire year of 1988. The new system is attracting a great deal of attention among art museum officials, both domestic and international. The same system has been introduced at the Saison Art Museum in Tokyo, and a new museum in a suburb of Tokyo (Machida-city) also recently decided to adopt the system.

HDTV has another potential application: generating tour guide information and ecological video pictures. The dynamic scenery of the Grand Canyon, captured by an HDTV camera attached to the front of a small jet plane, has no comparison whatsoever to any video images produced until now.

An attempt has been made to use closed-circuit television in combination with a communications satellite. In November, 1989, scenes of a Japan Cup Race, held at the Fuchu Horse Race Track in Tokyo, were taken with seven HDTV cameras, and the satellite was used to transmit the images to three different race tracks in Japan, where the viewers were able to see the Japan Cup Race on huge screens. The following December, a lively concert scene of one of the most popular Japanese rock music groups, filmed with HDTV cameras, was transmitted via communications satellite, and shown (at a charge) at five live houses around Japan. The plan proved to be a big success.

A gas company in Osaka recently installed two HDTV screens at a gas processing plant. One display is for showing sharp and detailed pictures of actual work on the operation line at the plant, while the other one displays computer graphics of the work process. Operators can supervise the entire operation by keeping their eyes on these two screens. An entirely new use for HDTV linked to a computer network was thus invented.

In the publishing area, attempts have been made to produce animal and plant encyclopedias, art encyclopedias, and general encyclopedias using CD-ROM or CD-1. HDTV image signals are being converted into digital numbers, then processed by computers to create photoengraving for printing. Several HDTV photography books made with this process are already on sale in book stores.

However, the area with the greatest potential for HDTV apart from broadcasting is probably the motion picture industry. HDTV will greatly reduce filming costs and allow for the creation of special effects with ease. This is not possible with conventional filming methods. HDTV technology has already been used, although only partially, in at least seven movie productions in Japan, including those of Akira Kurosawa. HDTV will also save on the cost of copying films, of which thousands of copies are made for commercial distribution. More importantly, however, HDTV will make possible the simultaneous transmission of movies via satellites and fiber optics, bringing about a revolution in the movie distribution system.

A plan for mini movie theaters incorporating the HDTV system is already within reach. While these theaters are now seriously being studied from various angles in Japan, a plan to link 14 HDTV theaters is already being realized in Florida.

We do not need to go through all the classic stories about historic innovations and discoveries to be convinced that the history of humanity's advancement would not have been possible without intellectual inspiration, supported by the goodwill and cooperation of a number of people.

In past years, there have been repeated discussions to determine an HDTV world standard. It appears, however, that these discussions have only led to an escalation of hostile debate, thwarting joint efforts to develop an advanced medium. HDTV, which is not a movie or a simple TV but a new medium, requires a pooling of different experiences and technical creativity in order to develop in the future.

# 5

# Moving Forward:
# Challenges and Opportunities

*Laurence J. Thorpe*
*Vice President, Production Technology,*
*Sony Advanced Systems*

*A*s a topic, HDTV looms large, and today embroils virtually the entire television industry, the larger electronics industry, and international administrations. As an issue, it can generate unusual levels of heated technical debate, political wrangling, confusion—even bitterness and fear.

But as an orderly and inevitable evolution to our existing and aging 525- and 625-line television systems, HDTV does not reflect the best of today's technological or entrepreneurial capabilities. It's a gigantic subject. The many implications of its arrival are only beginning to dawn on a uniquely diversified television industry and on the countless elements within a new information age who potentially can employ this powerful new electronic imaging system.

HDTV's advance, therefore, is a story of challenges and dilemmas that engulf many in the global communications industries. Objective recognition of these difficulties and the acknowledgment of priorities is central to finalizing the solutions that will open up opportunities never dreamed about.

No topic in television history has preoccupied so many for so long as has that of HDTV standards development in the 1980s. This is especially true in North

America. In the 1990s there are even more committees, splintered *ad hoc* and specialist groups and international liaisons all wrestling with a variety of issues that bear on HDTV/ATV standards. With the countless overlapping debates, there unfortunately emanate ever-widening confusion and dissension.

The dilemma here is that the topic is truly too large to be handled by any one single group. The number of technical variables combine with many viewpoints and special interests to generate a monster of debate within which painfully slow progress is made.

There persists a failure to adopt the macro view. The separate entities within the larger television industry regard each other with suspicion, and this has permeated all levels, technical and managerial. The challenge lies squarely with the governing bodies of the primary standards-making organizations. That challenge is one of management and, so far, it has not been well met. The unique method in the U.S. for generating industry standards revolves around a totally open forum of debate within the private sector, which produces voluntary-only standards. The Society for Motion Picture and Television Engineers (SMPTE), for example, has a fine history of standards making for production in both the film and broadcast television worlds based on this methodology.

But not so with HDTV. An unusually difficult and prolonged struggle ensues between aspiring future end users who set understandably high goals for a production standard and manufacturers who are all too familiar with the limitations of present-day technologies. The dearth of major professional television equipment manufacturers in North America with significant vested interest in HDTV compounds a difficult problem.

The current dilemma centers around a difficulty in absorbing the sober reality that good standards should be written about what can truly be manufactured in the near term. Only then will manufacturers be encouraged to make the substantial commitment of resources to produce equipment to that standard. Standards set "wheels" in motion for end users and for manufacturers. The same well-formatted standard should, however, include a carefully crafted blueprint that shapes a pathway to the future, allowing a sensible evolution of that standard to facilitate performance enhancements that are in step with downstream technological advances. Such an approach can only be rendered effective by an honest exchange between the primary factions within standards-related committees.

The challenges and dilemmas that permeate virtually all HDTV committees are caused by participants' struggles with conflicting feelings about specific application desires, bounds of technology, and an unavoidable level of politics and special interests. The very structure of SMPTE's current HDTV committees ensures a fostering of partisanship and a perpetuation of parochial in-fighting. It thus becomes a management challenge.

The opportunity lies in consolidation. Rather than a continuous splintering into new working groups, which in turn spawn new *ad hoc* groups, there should be a rallying of technical resources into more compact, focused, and tightly managed committees. The fact that this method can work has been well demonstrated by the *ad hoc* group on Digital Representation of SMPTE-240M (the current HDTV 1125/60 production standard), which pulled together a collection of technical talent from virtually all of the television disciplines. Remarkable accomplishments were achieved when the inevitable struggling gave way to the collective wisdom, which saw only advantages to a digital standard that encompassed the interests of all.

The Advanced Television Systems Committee (ATSC) also rallied to this challenge. An energetic new chairman set an agenda paced by international events occurring within the International Radio Consultative Committee (CCIR) study program on a possible international HDTV unified production standard. ATSC is a gigantic organization in terms of the breadth of the representation of its member organizations. Some exceedingly difficult discussions gave way to a consensus. Most of the topics were dealt with by a few specialist groups and a parent Technology Group. The executive committee defined tasks and timetables, and a vital point was proven: The task may be onerous, but it is manageable, and the results are so worthwhile.

The cost of HDTV is central to the larger discourse on the future progress of this new television system. The three separate portions of that total system each have a considerable stake in the overall economic picture: HDTV production, transmission/distribution, and display. Little can be said about the latter two, as no consumer HDTV products have yet materialized, nor can they until the technical problems of distribution media and their associated standards have been wrestled with and ultimately overcome. Happily, that process is now underway, and R&D activities are rapidly becoming more focused.

Comments can, however, be directed at HDTV production equipment, as products (conforming to at least one standard—1125/60 SMPTE 240M) are today readily available in the marketplace. The current costs of these products are inordinately high. Unquestionably, these costs coupled with the present standards confusion are impeding a more rapid advance of HDTV production.

There are a variety of forces that ultimately bear on the costs of HDTV production equipment. They include standards, technology, competition, breadth of application, and manufacturing economies of scale.

These forces are all inextricably tied together. The intense scrutiny of technical parameters has dimmed the necessary awareness of attendant costs. Technical parameters directly influence system performance and the technologies that must, of necessity, be harnessed to reliably and consistently realize such performance. One simple example graphically illustrates this direct relationship.

The first saleable HDTV production equipment became available in 1984 and a few entrepreneurs soon invested and set up operational production/post-production facilities. An average calculation was made of the prices of the basic HDTV products supplied by seven international manufacturers to some five facilities. These prices were then compared to the current prices (in today's dollars) of their standard 525/625 studio products. The HDTV equipment averaged 2.7 times greater than the standard products. Five years later, following completion of the SMPTE 240M 1125/60 standard in 1988, a new, second-generation HDTV product line became available from all of these same manufacturers and others. The average selling prices are now 3.3 times greater than the standard equivalent broadcast equipment at today's prices.

These price increases directly reflect the substantial increase in the technical specifications inherent within the SMPTE 240M standard, which include a 30 percent increase in bandwidth required for distribution; a wider color gamut; a wider aspect ratio (ratio of picture width to height), and a narrower horizontal blanking width.

SMPTE 240M carefully developed such performance enhancements to reflect the consensus of film and television experts that technical performance is of paramount importance within an HDTV studio origination standard intended both for high-end production work and for broad applicability. But these enhancements came at a price, and highlight a classic dilemma in standards development, namely, in seeking the "best," where do you stop? Today, there are already those who dismiss SMPTE 240M as inadequate in performance for an HDTV system that is expected to last three or four decades. There have been vigorous efforts made to rewrite the basic HDTV production standard based on progressive scanning, even higher spatial resolutions (2,048 x 1,152 digital samples), and even wider color gamuts.

Physics is unrelenting in the obstacles it places before real-time camera imagers (tubes or CCDs), VTR heads and tape, optical disc and laser technologies. Costs climb very rapidly as these technologies are pushed to their boundaries.

Of course, these costs will ultimately yield to competition and to manufacturing economies of scale. But the question is, how much? Today's production lines manufacture HDTV equipment in lots of perhaps ten. HDTV production equipment, however, like its 525/625 broadcast studio predecessor, will never escalate beyond production lots of hundreds. So costs will yield only grudgingly.

The solution offers an opportunity to those wise enough to grasp the significance of the problem. Develop a solid, sensible, pragmatic, implementable (with present technology) standard, and then maximize unification behind this standard. If the technology remains within bounds that can be addressed by many, and confidence is generated that one standard is acceptable to most, then (and only then) will true competitive manufacturing dynamics come into play that will inexorably drive down those costs.

There has never been a better opportunity presented for the entire television industry to unite to develop a standard equally applicable to future broadcast studio origination, motion picture production, computer graphics, and business and industry uses. Such unity would bring a vital new dimension to professional HDTV equipment economies of scale.

But if each segment of the television industry is to splinter and seek its own customized HDTV production standard, the results will be fatal. The cost increase between first- and second-generation HDTV production equipment signals a warning that should not be ignored. We simply don't have the luxury of writing exotic standards that speak to a possible global "leapfrog" into some ill-defined HDTV future that serves as a lure to manufacturers to escalate R&D resources to achieve only the ever-elusive "very best." The standard we need today is the one that encourages unity within a large and diverse professional television industry, and the one that is sensibly written to rally manufacturing commitment to an HDTV system that works well, now. Improvements to HDTV will come, and they will one day become cost effective, but they will be totally paced by key technologies and the rate and manner in which they improve.

It is worthwhile to examine another aspect of the challenges that lie before us as we explore what HDTV will mean to the future consumer; namely the advanced consumer display.

What will constitute the ATV screen of the future? A direct-view CRT picture tube or a projection system? This question quickly stimulates a complex discussion on aesthetics, ergonomics, cost, and picture performance. The dilemma exists because, as of 1990, we know very little about the quality of the ATV home imagery that will spark a major consumer turnover in television. We are immersed in complex technical discussions on what constitutes an HDTV production standard, with 35mm film hovering as the ever-immutable standard of reference. But what will American consumers "go for" when making what will surely be a significant buying decision? Woefully little has been done to seek a sensible assessment of consumer views on ATV imagery.

We do know, however, that the screen must be large. The CRT is worthy of scrutiny. The challenge here is to cost-effectively produce a large CRT with a 16:9 aspect ratio, requisite enhanced resolution (still to be defined), and adequate brightness. From a general design viewpoint our engineers list the basic CRT requirements as brighter and larger pictures, cost-effective production, compact depth, and compatibility with a current NTSC signal source.

Expanding some of these criteria into a possible specification for an HDTV CRT might produce a screen size with a 50-inch diagonal; brightness of 80 foot-lamberts; a contrast ratio of 100:1; more than 1.5 million pixels of resolution, a wide color gamut; 300-watt power consumption; a life greater than 15,000 hours; and a lightweight design.

Far more significant challenges lie, however, in realizing an advanced television (ATV) CRT that is sufficiently cost effective and of requisite size and weight to permit a really viable consumer product. The bulk of a CRT is difficult to diminish substantially.

CRT weight remains, perhaps, the most formidable challenge. As a large vacuum device, the CRT must be able to withstand significant atmospheric pressure—thus requiring a thick heavy glass structure. Sony's largest CRT available as of June, 1990, a 45-inch diagonal experimental unit, weighs 260 pounds.

The immediate alternative to the direct-view CRT is the CRT-based projection system, which employs optical techniques to enlarge small pictures. Remarkable progress has been made in professional HDTV rear-screen projection technology, for example, with 55-inch to 120-inch systems portraying very high brightness, contrast, and resolution. Nevertheless, the challenge of bulk and weight remain. The Sony HDIR 550—a 55-inch rear-screen 1125/60 HDTV projection system—has a depth of 35 inches and weighs 440 pounds. The challenge now lies in finding lower-cost, lighter-weight optics of a quality commensurate with future ATV consumer requirements.

Looking to the future, impressive progress is underway in liquid crystal display (LCD) technology. Plasma display technology (flat-panel video displays using ionized gas) is escalating as work increases on both AC and DC types, and the search widens to produce color plasma displays. Many technical challenges remain, however.

A great deal of speculation abounds regarding the relative merits of the HDTV image that might one day be presented to the consumer. In engineering circles, the technical parameters that constitute the multifaceted ATV image continue to be separately dissected with debates about the viewer's sensitivities to spatial resolution, colorimetry, grey scale, noise, temporal resolution and artifacts, and aspect ratio. On a more macro level, others assert that the impact of the wider screen with more resolution will have a lesser psychophysical effect on the average consumer than that of the advent of color over its black-and-white predecessor.

One of the dilemmas hovering over the global HDTV scrutiny is the lingering technical debate, which requires much-needed objectivity. Technical parameter values are hurled with gusto across countless committee tables accompanied by unique personal theories on ATV screen areas, shapes, brightness, and image content. Many demonstrations and tests have been conducted by well-meaning engineers, which appear to reinforce one view after another. But a significantly silent voice lurks somewhere in the background, that of the program producer.

There are not many producers who have had the opportunity to create imagery using this new medium. As a consequence, an important contribution to the general examination of HDTV is absent. A very key exploration has yet to take place, one that puts aside discourse on separate technical parameters and rather manipulates their aggregate imaging attributes to create pictures simply not possible within the severe restrictions of our present television systems.

The issue of the program producer, director, and cameraperson doing something different at the front end of the television system, the point of creation of the program, has not yet registered within current HDTV circles. And, it's a vital issue involving the implications of HDTV software, in contrast with those of hardware.

Psychophysics has shown that a good 525 NTSC image (portrayed on a current large TV set, say 27-inch) or a close-up camera shot, when viewed from our normal vantage in the living room, presents about as much information as our eye and brain can accept. But on a "long shot" the quality of the NTSC image collapses dramatically, the classic long shot of the football field being a good example.

If HDTV contains a total of six times more electronic picture information (as in SMPTE 240M, for example), then a screen size having six times the area of our reference 27-inch 525 NTSC screen would portray the same resolution per unit area (screen and viewer being the same distance apart). This could be exploited by presenting a much larger image of the picture being portrayed on the NTSC screen (with some 30 percent extra picture information on the sides due to the wider 16:9 aspect ratio). Thus a bigger picture with some more information is presented. Undoubtedly, this would have some impact: The picture would be wider and bigger with no loss of "unit area" quality. Overall, however, it represents merely a bigger version of the NTSC portrayal.

But the creative producer could also take note that a dramatically larger screen is available to portray another image he might choose to originate (with no loss of technical quality). Also, the viewing angle of the camera lens could be opened proportionally, to the degree that the center portion of the picture (the 27-inch diagonal center cut) becomes identical to that of the 27-

inch 525 receiver, but with the dramatic difference that a very large amount of additional picture information is added. (See Figure 5-1.)

**Figure 5-1**

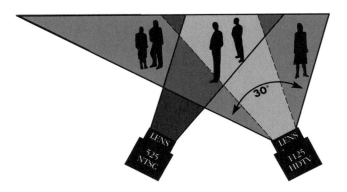

Wider, higher-definition pictures allow for more visual information to be conveyed on the screen.

Now an entirely new picture environment is created. Instead of merely packing more resolution into a screen, the screen and picture content both grow in direct proportion to the increase of information contained in the HDTV system. A whole new image emerges presenting a far greater sensation of reality.

It is proposed that when this important program software aspect becomes better understood, a new dimension will be added to the debate. The dilemma and challenge of presenting HDTV imagery to the consumer will lower dramatically. The viewer will see a far more dramatic portrayal of reality than that viewed through the necessarily confining technical "window" of 525 NTSC (or 625 PAL/SECAM outside the U.S.).

If this new creativity is applied to HDTV sports coverage, major events, drama, documentaries, and commercials, then a whole new "consumer debate" opens up.

The effect of this new imagery will be more profound than the addition of color to a still severely resolution-limited system of four decades ago. It is also suggested that the new form of television coverage unleashed by a new understanding of picture capture will lower the number of HDTV cameras required to cover many events, with a consequent cost benefit.

Finally, these new imaging possibilities have other possible implications on how our final programs are assembled. Today all the dexterities of electronic image manipulation are applied to stimulate the viewer. This artificial form of stimulation has evolved over the years as program producers instinctively sought to overcome the highly confining image presentation in our living rooms. When a dramatic new imaging possibility arrives in the form of large, wide-angle, widescreen pictures accompanied by multichannel CD-quality audio, it's very likely that production techniques will require radical new exploration.

The implications of HDTV program software are so immense that it is important that broadcasters and program producers begin experimenting now to explore all ramifications of this new television imagery. The opportunity is how to contribute an important new perspective to the overall discussion.

The advance of HDTV is inexorable. Only the pace is ultimately in question. What are at stake, however, are the opportunities that could be lost. These opportunities are immense when the broader view of HDTV is adopted. Of all that has been discussed here, the costs of HDTV represent, by far, the greatest challenge. These costs are a problem for all, and should be addressed by the collective efforts of all who would be involved in any way with HDTV. This calls for a unification of all television-related industries behind one production standard. Only by writing a pragmatic production standard that serves most of the needs of all, and by capitalizing on the economics of scale inherent in the use of that standard by all, can the requisite level of competitive energies by rallied to steadily lower those costs.

Once the system begins to function and HDTV starts

to be applied across a broad front of electronic imaging, then normal technological progress and continuing competitive dynamics will ensure an ongoing advance in the performance and capabilities of this new medium.

# ECONOMIC IMPLICATIONS

# 6

# Economic and Policy Considerations of High-Definition Television

*Dr. Corey Carbonara*
*New Video Technologies Project Director,*
*Baylor University*

*T*he evolution of new video technologies centers around the fact that whenever one focuses
on the development of technology, change is
inevitable, and whenever one focuses on the diffusion
of new video technologies, adoption becomes paramount. The process of technological change and the
spread of new video innovations can be currently
illustrated through the introduction of high-definition
and advanced television systems.

Charles G. Schott, III, former deputy assistant secretary of commerce, and the National Telecommunications and Information Administration (NTIA) identified
in 1988 four major areas where HDTV could make an
enormous economic impact. One of the areas is international competitiveness, especially when considering
HDTV's relationship to various industries, potentially
causing "ripple economics" to take effect. Secondly,
HDTV will have an impact on international trade; in fact,
HDTV may be the focal point of tomorrow's consumer-
electronics business. The NTIA values HDTV at $100
billion, representing new potential markets and a possible route to revitalizing the U.S. consumer-electronics
industry.

The third major area is communications policy.

Both the broadcast industry and the public have enormous stakes in the current NTSC system. Given the wide spectrum of alternate media, local broadcasting may be jeopardized if HDTV is not made available to terrestrial broadcasters. Video markets become most competitive when each media has access to new technology; therefore any rulemaking by the Federal Communications Commission (FCC) regarding HDTV broadcast transmission standards must take into consideration spectrum efficiency and compatibility issues. Finally, the area of spectrum management will be affected by HDTV as well. Overall responsibility for spectrum management lies with the NTIA and the FCC. Proposals to increase spectrum allocations have broad implications on other services and on U.S. competitiveness.

HDTV represents a synergy of technologies that are essential to the future of the consumer-electronics market, the telecommunication market, and computer markets. For example, suppliers of HDTV production, distribution, and receiving products and services will undoubtedly increase demand for semiconductors, thereby affecting that industry as well. The technological change brought on by the advent of HDTV has the potential to change the scope of technological leadership, ultimately affecting the political and economic leadership of those countries choosing to participate as players in the HDTV arena.

The development of HDTV has come about as a direct result of the Japanese concept of cultivating interrelated end-use markets, resulting in a series of interrelationships between products and markets, as well as the economic strength to support developing industries. For example, through the development of the videocassette recorder (VCR), the Japanese have consumed 12 percent of the total Japanese semiconductor output and 5 percent of the total global output.[1]

Estimates suggest that the consumer market for HDTV will reach as high as $35 billion annually in the U.S. by the year 2000. HDTV has a plethora of alternative media by which to enter the U.S. market. Cable, videocassette recorders, video discs, terrestrial broadcasting, direct broadcast satellites (DBS), and fiber optics are all staking their claims as the first to introduce HDTV to the American consumer.

Historically, the goal of a high-definition television system is to become the equivalent of projected 35mm motion-picture film; for over 100 years, television has aimed at this standard benchmark. Such a system would have a wider field of view (aspect ratio) and have enough resolution so that scanning lines would not be visible even at very close distances (three times picture height), thus enabling viewers to sit closer to the screen than they did before.

The features of an HDTV system involve a variety of fixed improvements over today's color television systems. HDTV would offer higher resolution (twice the horizontal and vertical resolution and five times the number of pixels of the NTSC system currently used in the U.S.); a greater aspect ratio than the 4:3 of current systems (for example, 16:9 has commonly been proposed); improved color, with at least 10 times the color information detail of present color systems; and improved audio fidelity, with multichannel sound capabilities equivalent to compact disc quality.

The existence of almost 200 million NTSC receivers and over 600 million current television receivers worldwide, however, has become a heavy consideration in the strategic implementation of HDTV. Those systems that do not adhere to any specific parameters of an HDTV system for transmission (such as 1125/60 or 1250/50), but which demonstrate improved quality over current systems, are identified as advanced television (ATV) systems. Recently, there had been three main categories of HDTV or ATV systems under consideration in the United States for transmission. They were: 6 MHz/NTSC compatible; 6 MHz/NTSC incompatible; and greater than 6 MHz/NTSC incompatible.

When the FCC began serious investigation of HDTV/ATV transmission, there were almost 20 ATV systems under consideration. On September 1, 1988, however, the FCC narrowed the field significantly by tentatively setting guidelines for establishing a terrestrial broadcast transmission standard for HDTV.[2] The guidelines stated that HDTV must be compatible with existing NTSC service; that no additional spectrum could be allocated outside the VHF and UHF bands; and that an inquiry be conducted on the relative advantage of a variety of allocation schemes for HDTV transmission. These schemes include those with one 6-MHz channel, those with one additional 3-MHz augmentation channel (not necessarily contiguous to the main channel), and those with one additional 6-MHz channel (not necessarily contiguous) as either an augmentation channel or a simulcast channel, during a transition period.

Through these tentative decisions, the FCC eliminated certain transmission systems from consideration for standardization—those with a contiguous 9-MHz channel, which are incompatible with existing NTSC.

By the fall of 1988, HDTV-related activities had greatly intensified in the U.S. In October 1988, the Secretary of Commerce under the Reagan administration, William Verity, appointed an Advisory Committee on Advanced Television to study the potential impact of HDTV on U.S. industries. That same month, the American National Standards Institute (ANSI) accepted SMPTE 240M (the HDTV production standard approved by SMPTE) as an American national standard for studio production. In December 1988, the U.S. government would once again play an active role in the technological development of television. That month, the Defense Advanced Research Projects Agency (DARPA) announced its availability of grants totaling $30 million for two HDTV projects. One project involved the development of HDTV display devices. The other project involved the development of HDTV video signal/computer processors.

In March, 1989, North American countries would establish a strategy that appeared to be similar to that taken by the Europeans at Dubrovnik in 1986. The North American National Broadcasters Association (NANBA) recommended that the CCIR not adopt a worldwide HDTV studio standard at the conclusion of its present study period in 1990. NANBA also recommended to encourage the identification of a "common image format" and the continuation of testing terrestrial formats, in order to establish potential relationships between emission standards and studio standards.[3] Also in March, 1989, The Sixth World Conference of Broadcasting Unions recommended studies based on two different approaches toward the achievement of a worldwide production standard. These two approaches involved systems with a common data rate and systems with a common image format.[4] (Common data rate is based on a common sampling frequency for all images regardless of frame rate or structure;[5] common image format is a system on a standard active-image structure with the same number of active lines per frame and active pixels per line regardless of the actual frame rate.[6]) On a similar note, in April, 1989, an ANSI Appeals Board rescinded its earlier approval of the SMPTE 240M after a second appeal by Capital Cities/ABC. The 1125/60 system once given enormous support by the U.S. was becoming increasingly challenged by some of the broadcasting and cable industry groups.

A variety of U.S. government and industry proposals were introduced in the first half of 1989. The new Secretary of Commerce under George Bush, Robert Mosbacher, outlined a plan that would relax antitrust laws and change the capital gains tax to cultivate HDTV manufacturing efforts by American firms. By May, 1989, eight Congressional bills were introduced on the topic of HDTV (six House bills, two Senate bills) advocating some form of government aid for strategic alliances, antitrust relaxation, and national cooperative research. Other interinstitutional players began to become more involved as the HDTV issues broadened to include other industries such as the computer and semiconductor industries. By mid-1989, industry groups such as the American Electronics Association (AEA) introduced an HDTV business plan to the U.S. Congress. The AEA plan calls for Congress to provide $1.3 billion in grants, loans, and credits for the development of a consortium providing a strategic alliance between government and American institution to research, develop, and manufacture HDTV products.[7]

In June, 1989, DARPA awarded its first round of HDTV display technology development bids to Newco, Inc., Raychem Corporation, Texas Instruments, Inc., and

Projectavision. Photonics Technology, Inc. was awarded a DARPA contract on flat-screen display technology.[8] According to DARPA, more announcements regarding HDTV image processor awards were forthcoming. And an additional $50 million for DARPA's HDTV research project was recommended by a Congressional subcommittee working on the 1990 Defense Budget.[9]

In August, 1989, a new hearing was set for a third appeal on the ANSI HDTV issue, this time validating the SMPTE 240M 1125/60 HDTV production standard. The ANSI/SMPTE standard stamp is important to the manufacturers and users of HDTV production equipment because of the need for common signal parameters for different pieces of the production process. Cameras need to be able to match with HD videotape recorders, switchers, and other signal processors. If standards do not exist, the risk of technical obsolescence or incompatibility becomes even more acute.[10]

In March, 1990, FCC Chairman Alfred Sikes stated the exact goals for the FCC in the implementation of an HDTV transmission standard. The FCC's intent is to select a simulcast HDTV standard that would be compatible with the current 6-MHz channelization plan, but would utilize design principles that would not be limited to the constraints of existing NTSC technology and would be independent of NTSC. The target for the standardization selection by the FCC is the second quarter of 1993. With the intention of the FCC to standardize a simulcast system, the possibility of approving an augmentation system was eliminated. Augmentation systems represent substantial spectrum availability and utilization problems.[11] The FCC formalized what the marketplace was leaning toward in the development of simulcast approaches to HDTV terrestrial broadcast. Developers of simulcast approaches are Zenith, the David Sarnoff Research Center/North American Philips consortium and the Japan Broadcasting Corporation (NHK).[12]

The "augmentation" approach to terrestrial broadcasting of HDTV allowed for existing television channels to continue to broadcast NTSC signals on their 6-MHz allocations. However, an additional 3 MHz or 6 MHz would be assigned to each station for the transmission of additional information that served to augment the picture resolution of the main NTSC transmission. In addition, side-panel information could also be transmitted in the augmentation channel, providing a 16:9 aspect ratio. Existing sets would continue to receive NTSC, while new HDTV sets would combine the NTSC channel with the augmentation channel to receive the full HDTV transmission.

The "simulcast" approach also allows existing television channels to keep their existing channel allocations. However, each station would also be given an additional 6-MHz channel to transmit a bandwidth-compressed HDTV signal. Existing NTSC receivers would take the NTSC signal, while new HDTV sets would pick up the HDTV signal. Simulcast is deemed superior to augmentation by some experts due to the fact that simulcast makes more efficient utilization of the RF spectrum.[13] In addition, simulcast systems are not constrained by the limitations of various NTSC artifacts. However, if the testing of simulcast systems indicates that such a system is not feasible for broadcast, then the commission may still select an augmented version of an extended-definition system. One important concern of an HDTV service is making sure that existing NTSC channels are not interfered with.[14]

The commission also stated that it would not make a decision on extended-definition television (EDTV) until it reached its primary goal of standardizing an HDTV standard based on a simulcast system approach. Therefore, the FCC decision did not stifle the continued development of EDTV systems. Proponents of EDTV systems are Sarnoff/Philips, NHK, Faroudja Laboratories, Production Services, and the Massachusetts Institute of Technology (MIT).

Regarding the distinction between HDTV, EDTV, and improved-definition television (IDTV), the FCC defined each technology in accordance with industry-wide accepted definitions of these terms that were developed by the U.S. Advanced Television Systems Committee (ATSC) in 1989. As stated earlier, HDTV represents a signal that provides at least double the horizontal and vertical resolution of existing 525 NTSC on a 16:9 aspect ratio, increased color fidelity, and CD-equivalent sound quality. EDTV represents a signal that would provide enhanced pictures to the home, including widescreen to special EDTV receivers, while at the same time providing a picture that could be decoded by existing NTSC receivers. IDTV represents a signal that improves NTSC without changing the transmitted NTSC signal. This includes improvements such as line-doubling (a method of increasing the number of scan lines by displaying each original line twice, which gives the viewer greater image density and cleaner delineation of the edges of different colors)[15] and ghost-cancellation (the addition of an additional signal in order to cancel out unwanted portions—ghosts—of a broadcast image).[16] According to FCC Chairman Sikes, IDTV is an area that does not require action to be taken by the commission in order to implement improvements to NTSC.[17]

In analyzing the technological, economic, and political considerations of HDTV and ATV systems, a differentiation must be made between three HDTV sectors. That is, the studio/production sector, which includes the production and post-production of programs; the distribution/transmission sector, which deals with delivery of programs to the home; and the receiver/display sector, which involves home viewing of programs. It is important to understand the principal economic and policy considerations relating to the diffusion of HDTV technology upon each of these various sectors.

In the studio/production sector, the drive for a worldwide production standard has been tantamount in importance to other production issues. The proliferation of new video distribution technologies to the home has produced a climate of hypercompetition for audience market share, necessitating new strategies to capture those audiences. These strategies could include more attractive programs, more expensive programs, and more international marketing of programs to meet additional costs.

To address these issues, program producers will need to enlarge audiences by exporting programs, using coproduction joint ventures to create international appeal, and enhancing technical quality of programming to satisfy a full range of global opportunities.

Currently, a high-definition production standard does exist and has for quite some time—35mm film; in fact, 85 percent of all prime-time programs produced in the U.S. have been on 35mm film since 1957.[18] Film production costs have increased at an average rate of 16 percent per year because of increased talent, labor, and materials costs.[19] Given these economic constraints and the need to insure more appealing products in order to remain competitive, other costs of production need to be stabilized and/or reduced.

Concurrently, a wide enough distribution reach must be maintained to earn profitable returns on investment (ROIs). Hollywood produces over 1,700 hours of original prime-time programs for the three major television networks—ABC, CBS, and NBC—with costs ranging as high as $1.2 million to $1.5 million per episode.[20] A studio executive at MGM/UA TV states that almost every network program is now losing anywhere from $100,000 to $300,000; therefore, profits can presently come only from domestic and international syndication, where as much as 63 hours per day of U.S. programming is broadcast in Western Europe.[21]

For certain types of programming (i.e., special effects), using HDTV for electronic production greatly improves productivity and cost economy by reducing the time spent in production. A CBS study conducted by Rupert Stow[22] compared a single episode of a one-hour drama produced in both 35mm film and HDTV, and concluded that HDTV provided an overall cost savings of 15 percent in production of prime-time dramatic programming and a more than 30 percent cost advantage over 35mm film for music videos and commercials. Principal costs savings came from labor, material, film processing, reduced production time, and reduced post-production (editing) time. Stow also concluded that a one-hour drama shot in HDTV could reduce production schedules from seven to six days per episode and reduce post-production from nine to four person-weeks per episode. According to Stow, the HDTV post-production

57

advantage accounted for a 46 percent savings in editing. By reducing production costs, using HDTV also allows a larger share of residuals in syndication and "value-added" to advertising campaigns. HDTV offers direct conversion for release via electronic media, which is a highly significant factor since videocassette revenues now exceed cinema revenues.[23] A single worldwide standard for HDTV production would have a number of economic consequences. Quality programs could be created on a wide scale at a manageable cost and international exchange markets for programs would lead to greater distribution, so companies using HDTV would show a higher ROI. If a worldwide standard for HDTV production is not achieved, however, 35mm film would remain as the international medium for program exchange, *de facto* standards would be implemented instead, HDTV equipment manufacturers would face a fragmented marketplace and lesser economies of scale, and program producers would need to have many different conversion systems to accommodate transcoding requirements.

Policy considerations for HDTV production standards include the need for fixed specifications in North America. This has been accomplished through the American National Standards Institute (ANSI)/Society of Motion Picture and Television Engineers (SMPTE)- and Advanced Television Standards Committee (ATSC)-approved HDTV 1125/60 production standard; however, it should be pointed out that both the National Association of Broadcasters (NAB) and the Association of Maximum Service Telecasters (AMST) voted against the 1125/60 standard on the grounds that local broadcasters need to participate in the full range of possible ATV technologies.[24] And although HDTV producers have noted a significant picture-quality difference between 1050/59.94Hz and 1125/60Hz systems,[25] two broadcasters—NBC and ABC—have shown support for broadcast HDTV production systems based on the 1050/59.94 system.[26]

In addition, commercial, government, and industrial applications have expressed interest in using HDTV for various nonentertainment applications, such as in medical applications, retail telemerchandising, teleconferencing, industrial quality control, training and simulation, tactical displays, defense, and CAD/CAM applications. However, the sheer size of the U.S. consumer market dwarfs any of these market subsets. In fact, it has been pointed out that the Department of Defense (DoD) would have to spend over 1 percent of its budget each year over the next decade to approximate the economic activity generated by a 7 percent penetration of HDTV into U.S. households.[27]

Recently, Japanese investors have turned to U.S. production and acquisition possibilities. JVC has invested in a Hollywood production company and, most notably, Sony purchased Columbia Pictures. Japan's decision to invest in Hollywood and establish joint ventures in entertainment came about, in part, because of the need for additional programming to support DBS MUSE, its direct broadcast satellite HDTV transmission system, during its early years and the larger need to develop Japan's own pay-television industry.

Significant policy and economic considerations surround the HDTV distribution/transmission sector as well. This sector is comprised of some of the following elements: broadcast networks and network affiliates, independent TV stations, syndicated program producers, cable television systems and program services, home video software, home satellite television, satellite master antenna television (SMATV), and multichannel multipoint distribution service (MMDS). The primary stakeholders in this sector accounted for a combined revenue of over $40 billion in 1987, representing about .9 percent of the U.S. GNP.[28] The demand for ATV services will divert time and capital away from existing video entertainment business; therefore, the impact of HDTV/ATV on these businesses will depend on the technical standards that are adopted and the relative timeframe for adoption. Some alternative media suppliers may also perceive HDTV/ATV technology as a means of differentiation in a maturing video market.

Distribution media operate in distinct economic conditions, which affect technical considerations. Regulatory environment, structure of ownership, competitive environment, and cost structures affect each distribution industry. Technical considerations can in turn affect economic conditions; for example, a greater bandwidth for an HDTV/ATV system generally produces greater picture quality, but also creates a greater opportunity cost because of the spectrum that becomes lost for alternate uses.[29]

Opportunity costs will vary between different media. In broadcasting, opportunity cost will be high for markets where channels are fully occupied; similarly, in cable, opportunity cost will be high because of retooling costs in repeaters, amplifiers, and converters. On the other hand, in satellites and fiber optics, opportunity costs will be incremental rather than direct.

The policy considerations of HDTV/ATV distribution and transmission media primarily center around spectrum allocations. In fact, the primary driver leading to the formation of the FCC Advisory Committee on ATV systems was the impending spectrum issue involving terrestrial mobile communications, when eight major cities claimed that spectrum was too scarce to support terrestrial mobile communications. The Land Mobile Communication Council initiated its lobby for more spectrum on the grounds that services will saturate existing allocations in New York, Chicago, and Los Angeles by the 1990s.[30]

Another spectrum policy issue concerns the allocation of the 12-GHz band. The Satellite Broadcast and Communication Association (SBCA) petitioned the FCC stating that 12-GHz frequencies should be reserved for DBS to maximize HDTV utilization, and cited several reasons that 12 GHz was unsuited for terrestrial broadcast use.

Compatibility is another important issue for this sector. The FCC tentative decision on compatibility with 6-MHz NTSC could be economically advantageous for broadcasters who already have technical and economic infrastructures in place, and will be able to continue using existing broadcast service. Some of the risks to broadcasters associated with the compatibility issue, however, are that alternative media will not be bound by either 6-MHz allocations or NTSC compatibility and that future policy may not require a "must carry" provision to cable operators, thus negating the possible advantages that the FCC's tentative decision provided.

Table 6-1 profiles the various distribution media and outlines the significant economic impacts of HDTV/ATV implementation.

**TABLE 6-1**
**Economic Impacts of HDTV/ATV Distribution Media**

**Fiber Optics**
Drivers:   Competitive positioning
              Revenue opportunities

Issues:    Competition for implementing fiber into the home among cable industry, telephone companies, utility companies

Impacts:  Cost of fiber to local loop as high as $200 per subscriber
              Delivery of interactive HDTV by end of 1990s
              Delivery of digital HDTV program distribution
              No constraints of 6-MHz bandwidth limitation

**Terrestrial Broadcast**
Drivers:   Competitive strength maintenance
              Stop eroding revenues
              Gain new revenue opportunities

Issues:    Compatibility with existing service in 6-MHz single channel vs. multiple channel approaches
              Need to deliver highest possible picture quality
              More capital expenditures needed for
              *Continued*

wideband service (was as much as $3.5 million for augmentation channel transmission studio and satellite receiving costs, could be less for simulcast depending on method of transmission)

Impacts: Competitive strength by offering HDTV/ATV service through current infrastructure
Solid programming base already exists
Solid financial base pre-exists through advertising

**Satellite (DBS)**
Drivers: Competitive positioning
Revenue opportunities

Issues: Competition from fiber, videocassette/vid-eodiscs
Superstation compulsory licensing
Costly launch and start-up costs

Impacts: DBS has total U.S. coverage, multipoint distribution
No 6-MHz bandwidth constraints, 9-MHz standard
Benefits with either compatibility or incompatibility

**Cable Television**
Drivers: Competitive penetration needs to reach 70 percent of number of households with cable to surpass terrestrial broadcast market
Revenue expansion from advertising and pay-per-view

Issues: Expensive material cost to implement ($100,000 per mile in urban areas,

*Continued*

$300,000 per mile when laid underground)
Primary competition from VCRs, TVRO
New potential competitive threat from telephone companies
Dual carriage expensive

Impacts: Cable not bound by 6-MHz restrictions but compatibility with NTSC does benefit existing base
Rechanneling will improve competitiveness in HDTV/ATV service
Over 7,800 cable systems in U.S. already service 20,000 communities
Can offer improved picture quality over broadcast service

## Home Video

Drivers: Competitive penetration
Expanding revenue opportunities

Issues: Continuing problems in piracy, sales vs. rentals, flipping, commercial "zipping" and "zapping" ($2 billion in revenue lost to video piracy annually)

Impacts: Multibillion-dollar industry
Changed cinema distribution patterns to home to less than one year after theatrical release
Lowered prices through economies of scale
Found partners in Hollywood where studios now generate more revenue from home video than the box office
Unaffected by 6-MHz bandwidth issue, could use this fact to gain advantage over constraints of other competitive media

From an economic perspective, many products depend on related systems—VCRs and prerecorded videocassettes are two examples. Certain risks to these products would occur due to a decentralized policy, in other words, no setting of technical standards. The risk to consumers is that technical obsolescence could force additional investments beyond the perceived value of the enhanced service. One benefit of standardization is that component compatibility translates into reduced costs; however, standardization would be costly for the enormous variety of already-existing products and infrastructure; their incompatibility with the new standards could become a hindrance to adoption of new products. Policy issues of standardization need to address the impact of regulation upon production economies of scale; the costs of compatibility; and the costs of transcod-ability (transform) relative to incompatible standards.

The economic and policy considerations surrounding the third HDTV sector (receiver/display) involve consumer as well as microeconomic issues. As stated earlier, by the end of the decade some estimates place consumer spending on HDTV products as high as $35 billion annually around the world. Projections of initial prices of HDTV/ATV receiver equipment in Japan range from $2,500 to $5,000, depending on screen size and features,[31] and HDTV VCR prices in Japan have been estimated to have a $3,000 introductory price.

Prices by the time of introduction into the U.S. could decline as much as 50 percent because of a possible fast adoption pattern in Japan producing economies of scale. Still, estimates place HDTV/ATV receiver prices at anywhere from two to four times the costs of a comparatively priced current NTSC receiver, and HDTV VCRs at about one and one-half times the price of current high-end NTSC VCRs.

Consumer acceptance of HDTV/ATV will depend on the consumer's expected perception of differences in these products and services compared to the additional investment in them. Features such as wide screens, higher resolution in luminance (brightness detail) and chrominance (color signals), and digital stereo sound are variables that may affect demand, and the demand for HDTV/ATV could be accelerated by relatively lower incremental costs, higher perceived quality, availability of larger screen sets, and downward compatibility with NTSC. Demand could be impeded, however, if there is a severe lack of programming available, a lack of relative accessibility to HDTV/ATV services, and/or a lack of downward compatibility with NTSC.

There is a causal effect between price and rate of adoption. Prices must drop from the introductory cost of $3,000 if more positive projections are to be realized. Current trends in component technologies suggest price declines should happen over time. Price elasticity—a small percentage in price reduction yielding a larger percentage in number of units sold—could further magnify the value of sales associated with greater demand.

The costs versus the benefits, both direct and incremental, of HDTV/ATV to the consumer will affect policy in this area. And these policy considerations will in turn affect how fast consumers adopt HDTV/ATV systems. The FCC's insistence on compatibility could send signals of stability to those broadcasters and manufacturers who are considering conversion to HDTV. If broadcast stations reach a level of approximately 13.5 percent adoption, "bandwagon" effects should further increase HDTV adoption.[32]

In addition to the microeconomic issues affecting the various HDTV/ATV sectors, macroeconomic or trade issues also arise in regard to these new technologies. HDTV/ATV standards and technologies will have implications on both domestic and global economies, affecting economic activity, employment, contribution to GNP, and the balance of trade. Individual industries will feel the effects at the national economic level, of course, but will also feel "multiplier effects." For example, studies show[33] that every $1 value added in color TV manufacturing results in a further $1.80 value added in other U.S. industries. This translates into about 35,000 jobs created for every $1 billion of value added.[34] At these rates HDTV/ATV is forecasted by economist

Larry Darby in a recent study to create 35,000 jobs in 1998 and 240,000 jobs by 2003; for HD VCRs the figure is 100,000 new jobs by 1998. (Total employment in 1986 for U.S. consumer electronics was 63,000 jobs.) Presently in the U.S. consumer-electronics industry, most value added occurs overseas with very little growth in domestic sales occurring since 1977, according to FCC figures.[35] The potential impact of HDTV/ATV could result in a positive change in the U.S. balance of payments if U.S. consumer-electronics manufacturers could obtain a larger share of ATV total sales. Thus, employment and GNP will be directly impacted by the amount of HDTV/ATV imports. (The imports of TV receivers and VCRs amounted to a $6 billion trade deficit in 1986.)[36]

Rapid development of HDTV production and distribution equipment on a global scale would be in the interest of the U.S. production and program distribution industries. HDTV would lead to expanded software exports, also cutting into the U.S. trade imbalance. In addition, some U.S. firms have begun to develop market niches for HDTV production equipment manufacturing, which relies heavily on digital component software, thereby leading to advances in that industry as well. If HDTV/ATV is not domestically cultivated in the U.S., the trade deficit could rise due to an ample supply of product coming to America from abroad.

European countries are strategically allied in HDTV/ATV through their collaboration on the EUREKA project, which is based on an evolutionary approach from the EEC-approved MAC (Multiplexed Analog Components) standard. Some experts speculate that ATV may diffuse more rapidly in Western Europe than in the U.S., based upon its market potential. Estimates have been projected as high as $25 billion by 2003.[37] Certainly Japan has reaped benefits from HDTV/ATV, both from its long-term strategy and commitment to HDTV and from its appreciation of developing global markets. Japan has invested in HDTV to produce functional systems along a well-defined production schedule, allowing for a leveraged position globally.

In conclusion, the use of HDTV in the studio/

production sector would lead to enormous potential revenue for program producers and suppliers, especially when combined with a worldwide standard for the international distribution of programs and with the cost reductions in production. Past failures in achieving a worldwide standard—AM stereo, color and monochrome television, ½-inch VCR formats—have resulted in a costly decrease in potential ROI on media productions and distribution. Cost-sharing based on program exchange and international strategic alliances would arise from a potential worldwide standard; the lack of a worldwide standard for HDTV production, however, would result in higher costs and expensive international distribution strategies.

In the HDTV distribution/transmission sector, certain distribution technologies will have greater opportunity costs and incremental costs associated with HDTV/ATV system implementation than others will. Spectrum allocation constitutes the primary policy concern affecting the distribution and transmission of HDTV/ATV systems, but compatibility will also affect the economic potential for certain media. Fear of technical obsolescence remains a significant driver to terrestrial broadcasters and is reflected by regulatory policy; it is important to analyze the cost relationships of standards compatibility as it impacts each alternative media. Cost-benefit analysis can be used to examine and clarify economic and technical issues and their resulting economic impacts upon distribution media.

In the HDTV receiver/display sector, there are enormous potential stakes associated with the HDTV/ATV industry. The prospects for positive adoption rates for ATV are based on the attitudes of consumers regarding HDTV/ATV products and on solid regulatory policies, a healthy economy, an investment by U.S.-based firms to support a long-term effort in HDTV/ATV, and the historical rates of growth of other similar technologies.

On a macroeconomic level, HDTV will affect the U.S. domestic economy and international trade. Multiplier effects of HDTV on other industries must be taken into consideration in the formation of any policy.

International strategic alliances such as the European Eureka project reflect the dynamic changes in global competitiveness; collaborative efforts provide a synergy not only to participating international firms but to the allied cooperative growth of whole industries, such as semiconductors. Potential economies of scale in these industries are linked to high-volume demand, and such demand can be stimulated by an HDTV/ATV industry.

## Endnotes

1. Richard Elkus, "HDTV Address," HDTV and the Business of Television in the 1990's Seminar, Washington, D.C., 9 Sept. 1988. Mr. Elkus is co-chairman of the ATV Task Force for the AEA.

2. "FCC Writes a First Draft for HDTV," *Broadcasting* 115.10 (5 Sept. 1988): 32-34.

3. The NANBA members are ABC, CBS, CNN, NBC, PBS (all American); CBC and CTV (both Canadian); and Televisa (Mexican). See David Hack, "High-Definition Television," Congressional Research Service, CRS Issue Brief (Washington, D.C.: U.S. Library of Congress, June 5, 1989).

4. The World Conference is made up of nine international regional broadcasting unions.

5. "Progress in Atlanta," *HDTV Newsletter*, 4.10/11 (March/April, 1990):8. For more information on CDR and CIF see also Stanley Baron and Kerns Powers, "Common Image Format for International Program Exchange." Presentation given to the 131st Technical Conference of the Society of Motion Picture and Television Engineers (SMPTE), October 21-25, 1989. Los Angeles, CA.

6. "Progress in Atlanta," *HDTV Newsletter*, 4.10/11 (March/April, 1990):8. A new proposal was recently introduced in Atlanta at the recent meeting of CCIR IWP 11/6 by Swedish Television. The proposal was for a compromise between CDR and CIF called "Common Image Part," which includes a common array for both 1125 and 1250 systems, at a 72MHz frame rate. For more information on CDR and CIF see also Stanley Baron and Kerns Powers, "Common Image Format for International Program Exchange."

7. John Gatski, "Capitol Hill's HDTV Plans Draw Positive Reactions," *TV Technology* 7.7 (June 1989):1.

8. For more information on the DARPA HDTV initiative see "DARPA Allocates Money for HDTV," *TV Technology* 7.9 (Aug.

1989):3.
9. Chip Cavanaugh, "ATV Funding May Increase; DARPA Selects 1st Companies," *Television Broadcast* (July 1989):1,23.
10. For more information on this twist of events in the ANSI standards battle for SMPTE 240M see Alan Carter, "ANSI Sets Third Hearing Date for SMPTE 240M" *TV Technology* 7.9 (Aug. 1989):1.
11. "FCC to Take Simulcast Route to HDTV," *Broadcasting* (Mar. 26, 1990):38-40.
12. Alan Carter, "FCC: Broadcasters Will Simulcast HDTV," *TV Technology* 8.4 (April 1990):6.
13. The RF spectrum is the set of frequency bands that are allocated for radio and television communications within the electromagnetic spectrum.
14. David Hughes, "FCC Decides to Push for 'Pure' HDTV," *Television Broadcast* 13.4 (April 1990):1,55.
15. Line-doubling can also be achieved by methods such as delta modulation (doubling the lines of resolution by interpolating new scan lines through the averaging of information from the lines adjacent to it.) For more information on line-doubling see Hans Fantel, "Bringing the Television Image into Focus," *New York Times* (January 8, 1989):28.
16. "Technical Report for Ghost Cancelling Systems," NEC Technical Report, *Computers and Communications,* draft paper for the National Association of Broadcasters, April, 1990.
17. "FCC to Take Simulcast Route to HDTV," *Broadcasting* (Mar. 26, 1990):38-40.
18. See Joseph A. Flaherty, "Television: The Challenge of the Future," *Television Technology: A Look Toward the 21st Century,* ed. J. Friedman (White Plains, N.Y.: SMPTE, 1987) 193.
19. Flaherty, 1987, p. 193.
20. Flaherty, 1987, p. 194.
21. Flaherty, 1987, p. 195.
22. See Rupert Stow, "The Economics of High Definition Television," 1987. A 1989 CBS "Movie of the Week" called *The Littlest Victims,* which was entirely shot in the 1125/60 HDTV system, indicated that the network's projections of cost economics not costing more in HDTV proved correct. See *The Littlest Victims,* prod. Fern Field, CBS, 1989.
23. Flaherty, 1987, p. 195.
24. See E. Feldman, "Advanced Television Update," *Via Satellite* (March 1988); Donow, *HDTV: Planning for Action.*
25. "Bottom-line Impact of HDTV," *Broadcasting* 115.2 (19 Sept. 1988):75
26. "NBC Unveils New HDTV Standard," *Broadcasting* 115.16 (17 Oct. 1988):31.

27. For more details on economic activity of HD penetration, see Larry F. Darby, *Economic Potential of Advanced Television Products* (Washington, D.C.: NTIA, 7 April 1988).

28. Federal Communications Commission, MM Docket 87-268, 88-288, 37462. As quoted in Federal Communications Commission, *Economic Factors and Market Penetration: The Working Party 5 Report to the FCC Planning Subcommittee on Advanced Television Service* (Washington, D.C.: FCC, 9 May 1988). This document was supplied to me as a member of the FCC Advisory Group on Advanced Television Service.

29. FCC, 1988.

30. Donow, *HDTV: Planning for Action* 29-34. For more information on 6MHz NTSC-compatible systems see Kenneth Donow, "ATV/HDTV: Evolution, Revolution, or Both?" in *Many Roads Home: The New Electronic Pathways* (Washington, D.C.: NAB, April, 1988).

31. Donow, 1988a, p. 25.

32. For more information on the diffusion of innovations see Everett M. Rogers, *Diffusion of Innovations,* third edition (New York: The Free Press of Glencoe, 1983).

33. See Arthur D. Little, *Consumer Electronics: A $40 Billion American Industry* (Washington, D.C.: Electronic Industries Association/Consumer Electronics Group, April, 1985).

34. Larry F. Darby, *Economic Potential of Advanced Television Products* (Washington, D.C.: NTIA, 7 April 1988) 14-16. For more information on the diffusion of innovations see Everett M. Rogers, *Diffusion of Innovations,* third edition (New York: The Free Press of Glencoe, 1983).

35. FCC, 1988.

36. FCC, ATV/WP5, 1988.

37. Darby, 1988, p. 46. (Washington, D.C.: NAB, April, 1988).

# 7

# HDTV and the Financial Effects on Broadcasters

*Michael J. Sherlock*
*President, Operations & Technical Services,*
*National Broadcasting Company, Inc.*

*R*ecently the American public has been hearing a lot about high-definition television, through frequent reports by television, newspapers, and magazines. The reports have presented glowing descriptions of the technological marvels of HDTV, and have looked with eager anticipation to 1993, when the Federal Communications Commission (FCC) is expected to announce its choice of a new transmission standard for American broadcast television.

However, the choice of a new system does not necessarily mean that typical American homes will quickly have their old TV reception replaced by broadcast HDTV in 1993. There are quite a few hurdles to be overcome first and the key here, as usual, is cost: cost to the consumer and cost to the broadcaster.

The issue of cost to the consumer has been addressed partially by the FCC, when it decided that the new broadcast system must not make the existing population of 160 million television sets in American homes obsolete. According to FCC estimates, American consumers have already invested $80 billion in their present TV equipment, which the FCC believes should not be rendered useless by the adoption of an incompatible system. Therefore the new system will be required to provide for continued use of the old equipment. After the new broadcasts begin, the consumer will have the option of purchasing a new advanced television receiver, capable of extracting a much-improved picture from the new broadcast signal. However, the consumer will not be compelled to make this investment. Anyone willing to accept the present quality of broadcast television will be free to continue using his or her present TV set. In this way the FCC acted to protect the consumer's past investment in equipment, but its action still leaves open the question of how to mitigate the huge cost to the broadcaster of making the transition.

Considerations of cost will be a major factor in the determination of which system of advanced television (ATV) should be chosen. The introduction of a new ATV service will be confronted by the classical "chicken-and-egg" problem: On the one hand, programmers and broadcasters will hesitate to accept the huge cost of producing and broadcasting the new ATV programs until there is a large audience with new ATV sets capable of displaying the improved programs; but, on the other hand, the potential viewers will hesitate to buy expensive new ATV sets until the new programs are on the air. Solving this dilemma will be a crucial factor in the selection for a proposed new broadcast system. Lower costs would help make the solution less difficult.

The FCC decision to alleviate the consumer's costs by protecting the existing population of TV receivers has placed a severe burden on proponents of possible new broadcast signals, particularly when coupled with another decision by the FCC: that the broadcasts would be in the existing 6-MHz channels of the UHF and VHF bands. These decisions have narrowed the possible choices to two types of proposed systems: simulcast (simultaneous broadcast) HDTV systems, using two 6-MHz channels (one for the HDTV signal and one for the standard NTSC signal), and extended-definition television (EDTV) systems using only the current single 6-MHz channel. There are several proposed systems of each type, with different potential advantages and disadvantages, but development is still underway and it is too early to predict the likely outcome. The FCC has expressed a preference for a simulcast HDTV system, if it turns out to be feasible.

In the simulcast option, the old TV sets will be accommodated by a standard NTSC signal, broadcast in one of two 6-MHz channels. The other channel is able to achieve greater efficiency by the use of modern, improved signal design, and is slated to carry the simulcast HDTV program. However, there is still a question of how much improvement in picture quality can be achieved by this improved signal design in a 6-MHz channel. To attain full HDTV—with double the resolution of NTSC and in widescreen format—the simulcast signal would have to carry five or six times the information carried in present 6-MHz channels. This is a very challenging requirement. Debate continues on how much information can be packed into a 6-MHz channel. Will a reasonable approximation to HDTV be achievable? Simulcast proponents claim they will be able to compress a signal of "HDTV" quality into a single 6-MHz channel and, although this ability has not yet been demonstrated, there has been a gradual growth in optimism that it may indeed be possible.

However, picture quality and wide screens are not the only considerations in the choice of the new television system. The relative costs of different systems need to be compared. The choice of a new broadcast system for the next generation of American television must take into account the realities of television broadcasting as a *business*. For example, picking a superior technical system that does not provide for a realistic and economically viable transition in the marketplace is self-defeating. The goal is to arrive at a better television system, but the choice must provide for a sensible way for everyone to get there. In evaluating different systems we must keep in mind the different transition scenarios each would imply.

The transition to a simulcast system contains some steps that are very difficult, time consuming, and costly. First, there is the problem of providing each broadcaster with an additional 6-MHz channel in the crowded UHF and VHF bands. Attempts are being made to accomplish this by decreasing the geographical separation distance between stations using the same channel, but this solution imposes severe interference-immunity requirements on the new broadcast system. At present it is not clear that any of the proposed simulcast systems will have sufficient interference immunity to allow stations to have adequate broadcast coverage areas. We might arrive at the planned decision date with the unfortunate conclusion that simulcast systems produce superior pictures but are rendered impractical because of their very limited geographic range. Imagine the chagrin of local station general managers who find out they can only reach, say, half their current audience with their new and expensive HDTV channel!

Another problem in the transition scenario for a simulcast system is the long delay involved in allocating spectrum (channels) and implementing the system. Even if sufficient spectrum is found and allocated, the process of assigning channels to individual stations, and possible court battles over the assignments, will remain. Also, there will be the need for new transmitter and antenna sites, which will involve additional delays due to the acquisition of properties, construction, etc. All of this means years of delay before the improved broadcast, and during that time the broadcasters will have no advanced television to meet the competition from rival media, such as videocassettes, direct broadcast satellites (DBS), and cable TV, all of which will have the opportunity to be implemented faster and further erode the broadcasters' audience share.

Finally, the transition scenario for a simulcast system entails a period of very heavy investment in entirely new and very expensive equipment. The average local station would spend somewhere from $20 to $40 million for production and transmission equipment. A report for Westinghouse Broadcasting estimated the cost at $34 million per station. Furthermore, for a long time there would be very little return on the investment, since very few viewers would have new ATV receivers. Faced with this dilemma, local broadcasters would be hard-pressed to justify the increased costs.

Therefore, if a simulcast system is chosen in 1993, there would not be a sudden blossoming of a new age of television, with HDTV broadcasts sprouting all over the country. The transition to the new system would be very gradual, step by step, with each station moving at its own pace. Different markets would respond differently, each according to its resources and its own view of its audience and its competition.

The expected transition to simulcast HDTV would be similar to the recent, and ongoing, conversion to stereo television audio. This conversion started in 1984, when the FCC established a *de facto* standard. Now, six years later, somewhat less than half of the commercial stations in the United States have converted to stereo. In some markets virtually every television station is broadcasting stereo sound, while elsewhere the conversion is just beginning. The decision is made individually by each station, depending on what other stations in the market are doing, and depending on the financial payoffs of making such a change. But, the costs of conversion to stereo audio are tiny compared with the huge costs of conversion to HDTV, so we can anticipate a much longer transition period for HDTV.

A strong force in determining the transition pace for each station will be the pressure of competition. The competition will come not only from neighboring television broadcasters but also from cable, VCRs, and DBS. Furthermore, since the rival media can implement HDTV or EDTV quickly, without the need for FCC spectrum allotments and other regulatory delays, broadcasters are under pressure to find a way to shorten or avoid the long delay required in implementing simulcast HDTV.

This concern about leaving broadcast television in a prolonged hiatus with regard to HDTV, with rival media providing superior service, has led some broadcasters to favor quick implementation of an EDTV system, either as a transition system during the long period before implementing simulcast HDTV, or as a more permanent system if the hopes for simulcast HDTV turn out to have been overly optimistic.

In the EDTV option, the picture quality would be better than present NTSC, but not as good as HDTV. The picture shape might be kept the same as for NTSC, which has a 4:3 aspect ratio, or it might employ the 16:9 widescreen aspect ratio of proposed HDTV systems. In either case the EDTV signal would be broadcast only in the current single 6-MHz channel and could be compatible with NTSC, so the old TV sets would be accommodated by the same signal that provides the new sets with improved EDTV pictures. This option has a much lower cost to the broadcaster, since a second transmitter and second channel are not required. Furthermore, the equipment costs for the single channel would be significantly lower than the costs that would be necessary for a simulcast HDTV channel, and the equipment changes would be evolutionary from present-day equipment, providing for a smoother transition with a more gradually phased investment.

From the viewpoint of costs, these features of an EDTV system are very attractive to the broadcaster. On the other hand, there is debate about whether the picture quality delivered by an EDTV broadcast system will be adequate to meet the quality of rival media, particularly if a proposed simulcast system is successful in its hope of delivering HDTV quality.

Furthermore, even though EDTV could be implemented relatively quickly at a cost far less than the cost of implementing simulcast HDTV, it would still not make good economic sense for broadcasters to implement EDTV as an interim system if the optimistic hopes about simulcast turn out to be justified. If the tests of proposed systems show that a simulcast system is capable of delivering near-HDTV quality with sufficient interference immunity and adequate coverage area, then there would be less of an argument for EDTV.

However, even if the simulcast development were successful, there would still remain some significant concerns about implementing it. Unless display technology advances much faster than is generally predicted, the television displays available in the next decade at prices affordable for general home use will not be capable of showing a significant difference between EDTV and HDTV. Television receivers capable of showing the higher quality of HDTV would be so highly priced that they would be affordable by only a small segment of the consumer market. This returns the debate over EDTV and HDTV to the "chicken-and-egg" question of what is the incentive for broadcasters to convert to HDTV if there is only a small audience for the improved signals.

Still, it is possible that the predictions about display technology are unduly pessimistic. If there is a dramatic breakthrough in the next few years, making it possible to produce HDTV displays at prices acceptable in the general consumer market, then the case for HDTV will be strengthened.

Another factor that may further complicate the issue is the rapid rate of development of digital video technology. All the proposed ATV systems depend crucially on the power of advanced digital processing of the signal in the studio and in the receiver, but the trend toward adopting digital methods is not as definite for the broadcast signal. The proposed formats for the broadcast signal include all-analog, mixed-analog/digital, and all-digital. Only one proponent has submitted a proposal for an all-digital broadcast signal, but others are continuing to work on the possibility of developing such a system. This trend away from the previous all-analog broadcast format is a radical advance, made possible by the increasing power of digital methods. The advance to an all-digital broadcast signal would be desirable for many reasons, and in particular would harmonize nicely with the eventual delivery of television by digital fiber to the home. This desirability has led to the suggestion that the introduction of a new standard for broadcast television should await the development of a practical all-digital broadcast system.

However, no practical all-digital system has been demonstrated yet within the constraints imposed on a simulcast HDTV system in a 6-MHz channel, and some observers are not optimistic that such a broadcast system could be developed in time for the proposed 1993 decision date. On the other hand, one proponent has already proposed an all-digital system, and at least one other proponent is actively engaged in research on perfecting such a system. In view of the unexpectedly rapid advances in digital technology in the past decade, it would not be too surprising if the efforts to develop such an all-digital broadcast system were successful.

Even if an all-digital broadcast system were not demonstrated by the proposed 1993 decision date, the progress made in its development by then might be sufficient to justify a clear expectation of success within a few additional years. This would seriously call into question the wisdom of any 1993 choice of a very expensive simulcast HDTV system, if it were apparent that the system could be rendered obsolete within a few years. All these uncertainties would contribute to the possibility of additional delays before implementing a simulcast HDTV system, which would tend to give additional support to EDTV.

The FCC has recognized the possibility of further developments, and has wisely provided a window in 1992 to review the status. At that time it should be possible to resolve many of the present uncertainties about simulcast HDTV, implementation delays, the feasibility of all-digital broadcast, and the possible need for EDTV.

In summary, it can be seen that the transition to a new broadcast system for the next generation of American television is likely to be a difficult and challenging time for broadcasting as a business. The costs of the changeover to simulcast HDTV will be huge and these costs, along with other market factors, would preclude any rapid conversion. The most realistic prospect is for gradual, step-by-step implementation, varying from market to market and station to station. The pace of changeover by broadcasters will be influenced largely by the need to match the performance of rival media.

Throughout the industry, there is general hope and increasing optimism that a simulcast HDTV system will be possible, perhaps even an all-digital simulcast system. However, success will depend on finding a system that can provide near-HDTV quality with sufficient interference immunity, adequate coverage area, and manageable costs. If it takes too many years to achieve this goal, then broadcasters may look more favorably toward EDTV, with its quick implementation and relatively low cost, as an attractive alternative in meeting the competition from alternate media.

The decision that the FCC will make in 1993 looms as the most important single factor in determining what system will be used, both in the short term and the long term, in bringing advanced television to the American public. Until then, continued development of all possible and viable options of HDTV and EDTV will put the broadcast industry in the best position to ultimately offer the American public the highest quality television at the lowest possible price.

# 8

## Cable-Ready HDTV:
## Cable May Be the First
## to Deliver HDTV to the Home

*Jeffrey A. Krauss, Ph.D.*

W hatever HDTV format is chosen for the United States, it will be a system that is unique. Other countries are in the process of adopting advanced television formats solely for satellite broadcasting, but only the United States is planning to adopt the new format for terrestrial television broadcasting. The cable TV industry, if it picks an HDTV standard that is closely related to the TV broadcasting standard, therefore has a unique opportunity. It can enable the United States to have a television distribution system capable of delivering advanced television by both terrestrial broadcasting and cable TV.

In September 1988, the FCC took the first official step toward HDTV in the United States, announcing the first set of tentative decisions regarding it. The FCC has decided that HDTV television broadcasting is in the public interest, and it plans to adopt a standard format that all U.S. broadcasters can use.

Another preliminary decision made by the FCC deals with nonbroadcast media. For satellite TV, cable TV, and fiber-optic distribution of video, the FCC decided not to adopt a standard. The standard format that the FCC adopts will only apply to terrestrial TV broadcasting.

This freedom allows cable and satellite formats to be driven by marketplace considerations. They will, however, have to be compatible to some degree with the broadcast standard, in order for a consumer to be able to use the same TV receiver to display programming from all sources. But cable and satellite systems are not as spectrum-limited as terrestrial broadcasting, so they may be able to use different formats to take advantage of their available spectrum.

Spectrum availability is the most serious constraint for HDTV broadcasting. Broadcast TV stations today use 6 MHz of spectrum each. In all major cities, the broadcast TV spectrum is fully utilized. Adding any more television stations would create interference to existing stations. This spectrum congestion exists largely because the technology used in TV receivers today does a poor job of rejecting interference, and because today's broadcast TV signal format is not very robust against interference.

The additional TV spectrum needed to support HDTV signals will have to come from two places: New TV receivers will have to be designed to be more resistant to interference, and a new HDTV signal format will also have to be stronger against interfering signals. Even then, TV receivers will still be subject to interference from strong nearby stations when the desired station transmitter is far distant and therefore presents a weaker signal for reception.

Cable TV, on the other hand, is not as limited in spectrum availability as is TV broadcasting. Cable TV, being a "closed" system that does not radiate out into the atmosphere, can actually use the same frequencies that are used in the outside world for airplane communications, police communications, radar, and other purposes. And because all the cable TV signals are sent down the cable with the same signal strength, the "near-far" interference problems that plague broadcast television do not bother cable TV.

Thus, a cable TV system that today carries, say, 50 TV channels and has a total capacity of 6 x 50 = 300 MHz could, with moderate expense, double its capacity to 600 MHz and give each of the 50 channels an additional

6 MHz of channel capacity. In contrast, it is not at all clear that the FCC will be able to assign an additional 6 MHz of spectrum to each of the broadcast TV stations now serving New York, Los Angeles, Chicago, or other major markets. The TV broadcast industry is fully aware of this disadvantage, and is scared to death.

As mentioned before, the FCC has said that it won't decide the question of HDTV standards for satellite TV, cable TV, or fiber distribution of video. These industry segments will have the freedom and flexibility to choose other formats.

This situation is not much different from today's, where satellite TV uses a different modulation (FM rather than AM) and a different audio format (digital sound in the horizontal blanking interval rather than analog sound as a subcarrier on the video) than broadcast TV. But today, satellite TV can be viewed on the same TV set as broadcast TV can.

If HDTV is to be a marketplace success, it is important that all of these video distribution services continue to be able to use a common display device. That means that whatever formats are eventually adopted by the cable TV and satellite TV industries need to be "friendly" and interoperable with the terrestrial standard.

For the most part, issues related to TV receivers and displays are being handled within the Electronic Industries Association (EIA), an association of equipment manufacturers. The EIA has set up a subcommittee to develop a family of TV receiver architectures for HDTV that can be compatible both with terrestrial broadcast TV and nonbroadcast media such as cable TV, satellite TV, and fiber. This compatibility is, in fact, required by the 1962 All Channel Receiver Act.

Any additional receiver capability will be an option that the manufacturer can choose in response to marketplace needs. Thus, we may see satellite-ready or digital fiber–ready TV sets in the future, in the same way that most electronic tuner TV sets are cable-ready today. And hopefully, the EIA subcommittee can decide on a simplified interface so that VCRs, cable TV descramblers, and satellite descramblers can be connected to the TV receiver with a minimum of user confusion and without signal degradation.

Because more than half of U.S. homes already receive their TV service from cable TV, "friendliness" to cable TV is an important decisional criteria for a TV broadcasting format. The broadcast industry knows that it would be shooting itself in the foot if it were to support an HDTV format that was totally incompatible with cable TV. Consequently, the testing of HDTV formats will include cable TV testing as well as testing on TV broadcasting frequencies.

The satellite and cable industries are potentially the big winners in HDTV. They have more technical freedom and flexibility than broadcast television. They will have HDTV programming available sooner than terrestrial television will, since pay TV programmers like HBO are likely to be among the first HDTV programmers. And, unlike broadcast TV, they won't have to wait until 1993 or 1994 for an FCC standard if they can reach agreement on an industry standard earlier. But there are still some unanswered technical questions.

The major technical question mark for cable TV is the effect on HDTV of the echoes and signal reflections in the cable. Cable TV systems have "close-in" echoes, which may be caused by loose or improperly installed connectors on the cable, producing ghosts that are very slightly separated from the actual image on the screen. The result is a softening of the picture sharpness that is usually not objectionable to the viewer.

Many of the proposed HDTV systems, however, achieve their higher definition levels and wider aspect ratios by using time compression at the encoder and time expansion at the receiver. It remains to be seen what impact these ghosts will have on the time compression techniques used by some HDTV formats.

Multipath reflections in video signals also produce "ghosts" in today's television signals. The time delay between the direct signal and the reflected signal determines how impaired the picture quality seems. If the reflected signal arrives very shortly after the direct signal, the ghost appears so close to the main signal that it simply causes a blurring of the picture rather than a distinct ghost. For echo delays of greater duration, the ghost can be distinguished separately on the screen. Of course, the wider the effective video bandwidth and the stronger the reflected signal, the more likely it is to cause an objectionable impairment.

With time-compressed video, as in some HDTV formats, multipath reflections give rise to two concerns. First, the time compression in the transmitted signal requires an expansion in the receiver; this expansion results in a longer effective time delay. Therefore the ghost appears farther away from the main image, and is more objectionable. Second, in HDTV systems that time-compress the luminance and chrominance components separately with different compression ratios, the result is two separate ghosts, one associated with the delayed luminance signal and one with the delayed chrominance signal. In contrast, with today's signals the entire signal is reflected and a reflection appears as a single ghost.

The concern over ghosts is based on theoretical analyses of the signal formats. Initial demonstrations of the Japanese E-MUSE equipment (which was designed for satellite broadcasting) on four U.S. cable systems during 1989, however, seem to indicate that HDTV signals can pass down a cable system without significant impairment. Of course, demonstrations are not enough. Testing of theoretical analyses under controlled conditions will be needed before any decision can be made.

Ghosting could be a serious problem for broadcast HDTV as well. Perhaps an approach for "ghostbusting" or echo cancellation can be found that can serve as an industry standard for both the broadcast and cable TV segments of the industry.

Meanwhile, municipal franchising authorities throughout the country are applying political and community pressure for cable TV systems to expand the number of channels they offer. For the most part, upgrading capacity means "rebuilds," that is, total or partial replacement of equipment and cables. But some cable operators are delaying their rebuilds until they know what performance specifications will be needed to support HDTV. This slowdown has caused some problems for cable TV equipment manufacturers, who risk dropping sales while the marketplace awaits direction. Cable TV systems are unwilling to spend money on new equipment if HDTV technology will make their systems obsolete in the next few years.

At the same time, HDTV is affecting the choice of the preferred cable trunking technology of the future. The cable TV industry has begun to show an enthusiasm for analog fiber optics, in spite of telephone industry claims that digital fiber to the home will have enormous capacity and eliminate the need for separate telephone and cable TV systems.

The cable industry appears to have already begun the evolutionary upgrade of its distribution plant using analog optical fiber systems, in part because the telephone industry's ability to carry digital HDTV on digital optical fiber is at a more rudimentary stage than over-the-air broadcasting of HDTV. The dozen or so HDTV transmission formats that are being evaluated by the FCC Advisory Committee are really compression formats; they were developed because an uncompressed HDTV signal would need a 30MHz channel, and there just isn't enough TV broadcast spectrum for this. On the other hand, the digital compression formats for HDTV supported by the telephone industry are still on the drawing board.

An uncompressed digitally encoded HDTV signal would require a data rate of about one gigabit per second, which is far too high for delivery to the home. Consequently, R&D work is needed to develop a digital compression scheme that allows the digital HDTV signal to be carried at a data rate of 130 megabits per second, or lower. In fact, no specific digital HDTV format has been proposed to date.

If the cable TV industry can directly use the broadcast HDTV signal ultimately chosen by the FCC, or adapt it for cable transmission with minor changes, it need not wait for digital R&D. By adding analog fiber into the cable plant, first on the trunk routes and then on the feeder and distribution lines, the cable TV industry can evolve to a fiber-optic design without the huge front-end start-up costs that will confront the telephone industry.

And the enormous capacity of fiber optics will allow cable TV to upgrade all of its programming to HDTV, while the spectrum limitations of TV broadcasting might forever saddle some TV stations with today's TV technology.

All in all, the cable industry is in a unique and attractive position for HDTV. Although many cable companies and manufacturers are working with the FCC and other agencies as they move toward decisions on HDTV, the decisions will have little or no negative effect on cable, regardless of what is decided.

The cable industry has been working on HDTV for years. And when there are HDTV signals ready for home delivery, cable will be ready to deliver them.

# 9

# HDTV and Fiber Optics

*Dr. Paul Polishuk*
*President,*
*IGI Consulting, Inc.*[1]

*B*y the middle of the 21st century consumers and business people will have available to them a wide choice of video services ranging from video telephony and "video mail" to dial-up video entertainment services. The media critic Marshall McLuhan has prophesied that the culture of the 21st century will be shaped and dominated by video communications. McLuhan sees both positive and negative aspects of that prophesy, but if it is to happen at all, it will be because of the evolution of broadband networks of sufficient capacity to carry multiple video streams and of sufficient sophistication to provide advanced video services.

The networks, organizations, and technology that will provide these services are already beginning to take shape as of 1990. But we are at a very early stage of their evolution. Nevertheless, it is already possible to see how the broadband networks of the future will emerge from a hotchpotch of new technologies and standards, which at present have little to do with each other. Of critical importance in this regard are fiber optics and high-definition television (HDTV).

The advent of HDTV, which will radically improve on the quality of video being delivered to homes, offices, factories, and places of learning, will do much to bring about McLuhan's vision. It will be critical in enhancing the video environment, bringing lifelike video to the home and office, and making video services more pleasant to use and easier to sell. And, in the long run, fiber optics is the only transmission technology that could (either theoretically or in practical terms) provide the kind of bandwidth necessary to make a wide variety of high-definition video communications services available to an extensive business and residential market.

But although fiber optics is now the medium of choice for point-to-point and long-haul communications, it is just coming into its own as a transmission medium for the last mile. Nevertheless, the demand for high-definition video services will fuel the demand for networks capable of supporting such services, which largely means fiber-optic networks. The construction of such networks will make it easier and cheaper to deliver high-quality video and will ultimately fuel demand for HDTV services. Thus the HDTV market will feed the fiber market and vice versa.

There are now a score or so of proposals for HDTV systems emanating from diverse groups ranging from major broadcasters to small research labs and universities. These proposals run the gamut from those that fit easily within the NTSC format currently used in the U.S., as well as within the present frequency allocations, to those that are completely incompatible with existing practice. NTSC compatibility is being required by the FCC for the HDTV standards that it will mandate for terrestrially broadcast television. These standards will also require that any additional spectrum for terrestrially broadcast HDTV will come from the existing spectrum allocation for broadcasting. The FCC standards for HDTV transmission may be decided on by 1992, but some observers suggest that a 1994 date is more likely.

NTSC compatibility in some form will also be attractive to cable television (CATV) and telephone companies because of the huge installed base of NTSC receivers. However, neither cable nor telephone companies have the bandwidth restrictions that apply to terrestrial broadcasters, and both will want to exploit this fact to deliver the highest-quality HDTV to those equipped with the necessary receivers. Beyond this, however, the strategic interests of the telcos and the CATV companies diverge. The cable companies will be looking for systems that fit within their existing analog networks, while the telcos are primarily interested in a digital future. Also because of the highly leveraged financial structures of many cable companies, these firms will be particularly attracted to any HDTV system that promises a low-cost implementation.

The digital-versus-analog issue has taken a back seat in much of the discussion about HDTV, but it probably will not do so for much longer. While it is unanimously agreed that HDTV will, in the first instance, be an analog service, it seems virtually inevitable that, in the long run, digital HDTV will be the norm. Digital television will offer the highest-quality image and sound possible and will also allow for the full integration of the television and the computer. Finally, digital HDTV will mesh smoothly with the digital formats that will undoubtedly dominate all forms of electronic and photonic (light-wave) communications in the 21st century.

But digital HDTV will have to await some improvements in compression technology to allow the signal to fit into a limited bandwidth (although the growth in fiber-optic networks should minimize the need for video compression). Even more importantly, it will have to wait for digital networks capable of supporting video services. The best hope for the emergence of such networks comes from the telephone companies' fiber-optic networks that are now being put in place. So far digital video standards do not exist except in the world of videoconferencing, where they do not address the issue of HDTV.

In the U.S., work on digital HDTV has been done at Zenith, General Instrument, MIT, and Columbia University.

Fiber optics has a number of characteristics that make it an ideal medium for all kinds of communications, but what makes it especially attractive for video transmission is its extremely high bandwidth. Fiber-delivered video is being rendered more practical by developments in the "microcosm," especially in the development of VLSI (very large scale integration, a chip technology) and integrated optics products that allow for signal processing and interfacing at the microchip level. Developments in these areas are expected to accelerate over the coming decade and will bring down considerably the cost of high-speed fiber-optic distribution of video signals. Improvements and cost reductions in light sources will also improve the chances for fiber-delivered video.

Fiber optics evolved primarily as a digital medium for voice and data communications. Although digital video over fiber has now been demonstrated many times, it has yet to prove cost effective for anything other than experimental installations. At some point in the mid-1990s, however, the declining cost of optical transmission technology and of broadband digital electronics, coupled with the widespread adoption of broadband standards, will allow digital fiber optics to prove its cost effectiveness for video.

In the meantime performance advances in light sources have allowed for the development of analog fiber-optic video transmission, which is now being widely adopted for the trunking sections of CATV networks. Various multiplexing schemes can be used to carry multiple signals over a fiber, but subcarrier multiplexing (SCM) is the only technique that allows both digital and analog signals to be mixed on a single fiber. This makes SCM an attractive interim solution prior to the arrival of digital video, since with SCM analog video can be mixed with digital video and/or audio. Cable companies can used SCM to provide CD-quality sound along with high-quality analog video. Telephone companies can add digital voice services, which will become the norm as Integrated Services Digital Network (ISDN), a digital network allowing the transmission of computer data, voice, still and moving pictures, and high-quality sound, penetrates during the 1990s.

Within the analog video world, FM schemes have taken up much of the market, but AM schemes are now being adopted. AM has the advantage that it is directly compatible with existing video terminal equipment so that the signal can be fed directly into televisions, VCRs, etc. FM signals must be processed. Although the FM processing adds costs to the entire system, however, the extra expense is not huge because of the high level of integration in the necessary circuitry. FM transport also yields a considerable "processing gain" that translates into improvements in link performance and improved video picture quality. It has therefore been suggested that, despite the need for processing and the fact that tuners and demodulators are required at each video appliance, the higher signal-to-noise ratio obtainable from FM may make it a more suitable technology than AM for HDTV.

For the next few years it seems likely that AM and FM schemes will battle it out in the marketplace, without either one being in the ascendant. Eventually, both will give way to digital video. Initially, this will employ SCM, but may evolve to using TDM (time-division multiplexing) when the necessary interfaces for video appliances are available.

In practical terms, video compression is likely to continue to be necessary in any future fiber-delivered video scheme, despite the high bandwidth available from fiber. It will be all the more necessary as HDTV services are introduced because of the bandwidth-hungry nature of HDTV. Video compression techniques are now more than adequate for dealing with fiber-delivered HDTV, but the economics of video compression are likely to improve as VLSI and integrated optics technology improves.

Error correction schemes are quite important with compressed video, because transmission errors have a greater impact on the signal quality at the receiver compared to uncompressed video. However, once again, technical means of overcoming this problem have now been developed.

A whole new generation of standards are evolving to meet the very special needs of fiber-optic networking in both the public carrier and private network environments. In the public arena the most important of these standards is broadband ISDN (B-ISDN), which will be the high-speed, fiber-delivered successor to the ISDN that is currently being implemented by the telcos. B-ISDN will also incorporate two other important fiber-optic standards/techniques. The first is called SONET and is a hierarchy of standard electro-optical interfaces. The second, ATM, is a high-speed "fast" packet switching technology.

Video delivery (both HDTV and non-HDTV) is at the core of the B-ISDN philosophy and the evolving standards make explicit a wide range of video services. B-ISDN also defines five quality levels for video. HDTV represents the highest service level in this hierarchy and the B-ISDN standards developers envisage that it will be encoded at between 92 and 200 megabits per second (Mbps). The implementation of B-ISDN will ensure a place for a wide range of HDTV services in the repertoire of public network communications services, but it will be well into the 21st century before the impact of B-ISDN makes itself felt outside of a few specialized circumstances.

In addition to fiber-optic public networks, standards making is also beginning to address the issue of fiber-optic local-area networks (LANs). LANs (fiber-optic or otherwise) are primarily data networks. There has been little concern with providing for video applications; two important exceptions, however, are the American National Standards Institute's (ANSI's) FDDI-II standard and the Institute of Electrical and Electronics Engineers' (IEEE's) 802.6 standard. Both of these standards would be suitable for digital HDTV transmissions, although this is not a primary part of their purpose. FDDI-II is a wholly fiber-optic standard, while fiber is the medium of choice for IEEE 802.6, which defines a LAN-like network for use in city-wide metropolitan area networks (MANs).

There is something of a conflict of visions when it comes to government policy on the development of fiber-optic networks. On the one hand, federal and state governments frequently claim (as they do with HDTV) that their policy is to encourage further fiberization. On the other hand, laws and regulations designed to address non-fiber-related issues seem sometimes to have created a rather restrictive regulatory environment for local fiberization.

Rules preventing telco competition with CATV

operators tend to have a dampening effect on the fiberization of the telco's local loops since much of the incentive for telephone companies to run fiber into the home is to provide video entertainment services. These restrictions on the telephone companies' rights and abilities to provide advanced services is likely to have a chilling effect on fiberization in general.

There are also many local regulatory restrictions on communications competition. Such restrictions allow telephone companies to continue to use older technology and prevent the construction of independent MANs, which are highly likely to use fiber optics.

Finally, Congress is threatening to "reregulate" the cable television industry. This would probably reduce the profitability of the CATV industry and make it less likely to invest in advanced technology including fiber.

In virtually every case, the regulatory restrictions and proposals that impede fiberization are under fire from various quarters. Increasingly, today's regulations look like they will not make much sense as a means of controlling tomorrow's technology.

Although "industrial policy"–type arguments for government funding of HDTV rest on the broad applicability of HDTV to consumer markets, there is relatively little discussion of HDTV outside of the residential sector; these nonresidential uses will be critical to the development of the whole technology. There is little doubt about the American public's significant ability to consume video entertainment services, but whether it will be willing to pay $2,000 or so to upgrade to HDTV remains to be seen.

The transition to HDTV in the residential sector will probably be quite slow and, initially, only a tiny proportion of households will be receiving HDTV signals over fiber, rather than through over-the-air broadcasts and coaxial cable provided by the CATV companies. However, as the telcos' switched-fiber-to-the-home trials evolve into commercial networks, more and more HDTV will shift to this medium, especially as dial-a-movie and video library services are introduced.

In the commercial and industrial sector, the demand for videoconferencing services for business meetings and educational applications has been well established over the past decade. Future fiber video networks will go a long way to making such services easier (and perhaps cheaper) to access. The role of HDTV (over fiber or otherwise) is not entirely clear, but will certainly be implemented to make videoconferencing a more pleasant service to use.

Future fiber-based carriers may find that despite all the media fuss about fiber to the home, it is business video and HDTV services that generate much of the immediate demand for fiber optics.

Medical institutions use video services for similar applications to the business community. However, the medical community also needs networks to transmit high-quality imaging information. Fiber-delivered digital HDTV networks could serve very well in this way.

Educational institutions are likely to use switched-fiber video networks in long-distance learning applications. The introduction of HDTV will enhance the applications where visual imagery is important.

The major video services bought by military and government are the same as those purchased in the civilian sector. Although the importance of HDTV in defense is often emphasized by commentators, the development of high-resolution displays is of more importance in this context than is networking.

Commercial opportunities will flow from the synergy between fiber optics and HDTV. Many industries and markets will benefit from fiber optics' HDTV capabilities in the late 1990s and into the next century. Both business and commercial applications of HDTV will find opportunities in HDTV, from cable services and telcos, to computer and data networks, and extending to members of the film and television community.

In most areas, it is expected that trial services will begin in the 1993-to-1995 timeframe with full implementation to begin in 1995 and carry through the year 2000. It is probably not until the first decade of the 21st century that HDTV will become a standard video format, or that fiber optics will realize its full application potential.

## Endnote

1. Excerpted from the Executive Summary of *HDTV and Fiber Optics*, ©1990 Information Gatekeepers, Inc., 214 Harvard Avenue, Boston, MA 02134.

# 10

# HDTV's Relationship to the Semiconductor and Other Industries

*U.S. Congress,*
*Office of Technology Assessment*

*H*DTV is possible only through the intensive use of digital electronics; significantly higher-quality pictures than today's NTSC cannot be delivered to the home by any other means because of bandwidth constraints. As a result, the core technologies of HDTV—production, storage, transmission, processing, and display of information—are the same as those used in computer and telecommunications devices.

It is often overlooked in the current debate that HDTV is a development vehicle for "high-resolution systems" (HRS) generic to all information systems. HDTV proponents, for example, argue that the ability to produce high-performance displays and other technologies gained from a presence in the HDTV market will give manufacturers significant advantages in producing related components and systems for the computer and telecommunications markets. Skeptics resist the linkage argument as an unproven hypothesis and insist that these technologies can be developed by the computer and telecommunications industries independently.

The U.S. Congress' Office of Technology Assessment found evidence that HDTV developments *are* driving the state of the art in several of these technologies more rapidly than are developments in computer or telecommunication systems. The enormous amount of information in a real-time, full-color HDTV signal—some 1.2 billion bits per second (1.2 Gbps) in the uncompressed signal for 1125/60 HDTV—places severe demands on today's technologies. This contrasts sharply with the conventional stereotype of consumer electronics as low-technology products lagging far behind the leading edge of computers and telecommunications.

HDTVs must handle huge information flows and require special hardware to provide high computational speeds to convert a signal compressed for transmission back into a viewable picture. Digital signal processors (DSPs) tailored to this specific task are meeting that requirement. In contrast, engineering workstations, for example, must flexibly perform a broader range of calculations than an HDTV; therefore, they are software programmable.

The major technological bottleneck for the workstation today is the computational speed and the flexibility of its microprocessor and graphics display chips. Workstations put less stress on communications, storage, and certain aspects of display technologies than HDTV because they do not yet approach the information flows or the (specialized) computational speeds demanded of HDTVs.

Other high-resolution systems, such as desktop publishing and medical imaging, place different and often lower demands on DSP, storage, and communications technologies than does HDTV. They typically do not operate in real time and accordingly have lower rates of information flow. Many of these other applications also place lower demands on display technologies than does HDTV. They work well enough with slower response times, limited colors, lower brightness, or smaller display areas. Providing sharp images of stationary objects as is usually the case with computer applications is, in many respects, technologically easier than providing high-resolution, real-time, full-motion video. These other HRSs often require, however, higher resolutions than currently planned for HDTV displays. Because computer displays are normally viewed close up and there are eye-fatigue issues, some of the design criteria for non-HDTV high-resolution systems are different.

While many of the linkages between these technologies are obvious, they are not easily measurable. Nevertheless, these linkages can have an enormous impact on widely scattered technologies and markets. Simple analyses in which the projected future value of an industry is discounted to the present cannot account for the new and unforeseen opportunities that might be created by being in a market. Sony's and Philips's development of the compact disk player, for example, has opened a huge market in computer data storage. Similarly, flat-panel plasma, non-impact printers, and electroluminescent displays, among others, drove the initial development of power integrated circuits (power ICs), which, in turn, are revolutionizing the distribution and control of electric power in equipment ranging from aircraft to air-conditioners.

The linkage argument does have limitations. For example, unlike leading-edge PCs or workstations where performance is everything and price a secondary consideration, HDTVs must be produced and marketed at a price within reach of consumers. This demands exacting design and manufacturing discipline that is often lacking in narrower or more specialized markets, such as the military or medical imaging.

The large potential size of the HDTV market could enable significant improvements in manufacturing technology as firms seek to lower production costs. In some cases, low-cost manufacturing will require pushing the state of the art in component technologies; in others, it will mean that HDTV will let the computer or telecommunications markets push the state of the art and will then use those results. Above all, HDTV—as for all consumer electronics—will require pushing the limits of cost-effective manufacturing of sophisticated electronic systems. This might be one of the most important impacts of HDTV.

Whether or not a consumer HDTV market develops, the expectation that there will be a large market is forcing manufacturers who wish to participate to push the state of the art in the various HDTV-related technologies. If the market does develop, then large-volume production might give the producer economies of scale in a number of other components and products.

Having a technology matters little if markets are closed to innovators or the entry barriers are effectively insurmountable. As a result, it is also important for those that develop the technology to capture a significant share of the market. In the past, the United States has assumed that if the technology was developed, markets would follow. Faced with large, usually aggressive foreign competitors, and confronted with increasingly skill- and capital-intensive R&D and manufacturing to produce high-technology goods, this assumption is no longer valid.

Neither linkages nor market share nor volume production of computers were sufficient to save the U.S. DRAM (dynamic random access memory) business. U.S. firms produce about 70 percent of the world's personal computers today and lead the world in PC design. Nevertheless, domestic firms have lost the market for DRAMs to Japanese firms. A combination of factors, including less efficient manufacturing by some U.S. firms on one hand and aggressive foreign trade practices on the other, forced most U.S. manufacturers out of this important market.

The United States has also lost important HRS imaging markets that already exist, such as low-end copiers, as well as many other pieces of the electronics industry, despite having a predominant market share in many of these just a few years ago.

Start-ups in the U.S. electronics industry are increasingly focusing on design alone and depend on foreign operations for the highly capital-intensive manufacturing operations—they cannot secure the capital necessary to do the manufacturing themselves. In contrast, a number of foreign firms with little expertise in advanced electronics are becoming important manufacturers of electronics through heavy and long-term investments and careful attention to the manufacturing process. For example, NMB semiconductor, a new subsidiary of a Japanese ball-bearing company, Minebea, in just five years entered and became the world leader in very fast DRAMs. Kubota, a Japanese agricultural equipment company, is now manufacturing minisupercomputers designed in the United States. Similarly, Korean semiconductor firms are now becoming important producers of commodity DRAMs.

The United States cannot survive by performing R&D alone. Manufacturing provides far more jobs, greater value added, and larger cash flows than R&D, and these are needed if future investments are to be made in R&D or production. Nor can every American earn a living as a design engineer. With the increasingly tight linkages between R&D and manufacturing due to the exacting requirements of modern manufacturing processes and quality control, and due to the need to design for manufacturability, R&D is merging with the manufacturing process. In many cases, when the United States loses manufacturing, the loss of R&D is not far behind.

The problems facing the U.S. electronics industry are much broader than simply HDTV. Although HDTV may be an important element of any broader U.S. strategy in electronics, by itself HDTV will neither seal the fate nor save the U.S. electronics industry. There are undeniable linkages—some strong, some weak—that should be recognized, but the problems facing the U.S. industry extend into many other financial and structural factors. These include: the higher cost of capital in the United States resulting, in part, in lower capital and R&D investments than its competitors; inattention to manufacturing process and quality, poor design for manufacturability, and separation of R&D from manufacturing; foreign dumping and foreign market protection; and smallness of scale and/or lack of vertical/horizontal integration compared to foreign competitors.

The rapid technological advances and cost reductions of digital electronics will likely make HDTV affordable in the not-too-distant future. During the past 10 years, the capacity of leading-edge memory chips (DRAMs) has increased by 250 times while the cost per unit memory has decreased nearly 100 times. Each generation of advanced TVs will use increasingly complex digital semiconductors to provide a better-quality picture at a lower cost.

In turn, HDTV will directly push the state of the art in various aspects of DSP, display, data storage, and possibly semiconductor packaging technologies, among others. HDTV may indirectly impact a much broader range of components as well as computer and telecommunications systems as a result of these technological advancements.

There are three steps in digital processing. First, the continuously varying analog signals of the real world—sound, light, temperature, etc.—are converted into a digital form usable by computers with an analog-to-digital (A/D) converter. Second, the signal is processed with a digital signal processor—to decode the tightly compressed broadcast signal back into a recognizable picture or reduce ghosts and snow (noise) to produce a nearly flawless picture. Third, the digital signal is converted back into an analog form—sound and pictures, etc. that people can understand—with a digital-to-analog (D/A) converter.

A digital signal can be manipulated (as in signal compression), analyzed, transmitted with greater reliability, and stored in computer memory. In general, the more the broadcast signal is compressed to fit into a narrow bandwidth, the more digital signal processing power is required for its reconstruction. The development of better video signal compression algorithms is an important nonhardware aspect of current work in HDTV that is also critical to the success of interactive video systems.

Digital signal processing is used today in compact disk players, facsimile mail (fax), long-distance telephone lines, computer modems, and in other applications. Human hearing and vision are analog, so digital signal processing will play an increasingly important role in providing an "interface" between people and information systems in the future as we come to rely more on images and sound instead of alphanumeric text.

Digital signal processing chips (A/D, DSP, D/A) and digital signal processors in particular for HDTVs are at the leading edge of many aspects of the technology. For example, at the 1989 International Solid-State Circuits Conference (the most important international conference for unveiling new chip technologies), some of the fastest DSP chips ever developed were specifically designed for HDTV or related color video signal processing. The only comparably demanding applications today, at least in terms of speed and throughput, are military radar and sonar (highly specialized and low-volume markets) and image processing that is closely related to HDTV.

An uncompressed HDTV signal contains about 1.2 billion bits per second (1.2 Gbps) of information. In comparison, today's advanced engineering workstation hits peak speeds internally of roughly one-half gigabit per second. This signal is compressed, transmitted, and is then converted back into a viewable picture by the receiver. The amount of computation needed to decode this varies with the standard chosen, but can be as much as two to three billion mathematical operations per second. DSPs are able to handle these huge information flows and computational speeds—roughly comparable to those attained by today's supercomputers—at a cost consumers can pay, only through specialized designs tailored for specific tasks. Unlike HDTVs, supercomputers are able to handle a broad range of computations and to do so much more flexibly.

The development of certain important computer technologies may be aided in part through efforts in developing digital signal processing for HDTV. For example, the "testbeds" built by the Japanese to develop DSPs have required extensive work with massively parallel processor systems—the ability to hook up many microprocessors in parallel to speed up computations. At Nippon Telephone & Telegraph, for example, the National Academy of Sciences Panel reviewing Japanese HDTV development efforts observed a system with 1,024 processors in parallel, far fewer than some U.S. systems but still a notable achievement. Parallel processing has been a significant weakness in Japanese supercomputer technology, and a primary area in which U.S. firms have managed to maintain their edge. The experience with parallel processing hardware that the Japanese have gained in their HDTV development efforts may have spinoffs to their supercomputer systems.

Similarly, HDTV research at the David Sarnoff Research Center has led to the development of a video-supercomputer capable of an information flow rate of 1.4 Gbps and computational speeds of some 1.4 trillion mathematical operations per second at a cost of less than one-tenth that of other supercomputers.

The DSP market is growing rapidly. It is expected to increase from about $650 million per year to $1.6 billion by 1992. If the HDTV market develops, HDTVs will use an enormous amount of digital signal processing. For example, the Japanese MUSE-9 system uses some 500,000 gates—a measure of processing power—to convert the highly compressed signal back to a picture. DSP requirements may be less or more than this, however, depending on the eventual choice of transmission and receiver standards. (MUSE-9 is not being submitted for FCC consideration for a domestic U.S. transmission system.)

If high rates of growth are realized for the HDTV market in the United States, Japan, and Europe, then 15 years from now the use of digital signal processing chips in HDTVs alone could be 10 times today's total world demand (measured by processing capacity—"gates") for all microprocessor and related applications and roughly 100 times today's demand for DSP.

Such estimates are speculative. Depending on the relative growth of the computer, telecommunications, and other markets, this may or may not be important compared to the entire microprocessor and microcontroller market in 15 years. It would, however, almost certainly have a strong impact on the cost of DSPs for video processing. Even under low-growth scenarios a tenth as large, HDTV would likely have a strong impact on the cost of DSPs for video processing.

The United States currently has a stronger position in DSP design than Japan and is about equal to Japan in the production and performance of DSP chips. Domestic firms have managed to maintain a dominant market position in DSPs because they have better software for developing these chips. Texas Instruments currently has 60 percent of the world DSP market; NEC is second with 11 percent. Production of DSPs for HDTV could significantly change these market positions, depending on a firm's presence in HDTV. Such concerns may have been a factor in Texas Instruments' decision to purchase the Japanese HDTV chip designs and technology from Japan's NHK in order to participate in the Japanese HDTV market.

HDTV television receivers similarly place heavy requirements on memory technology. Access times needed for HDTV memory chips must be roughly 20 nanoseconds (ns)—20 billionths of a second. Today's fastest DRAMs have typical access times of 60 to 80 ns.

Leading-edge PCs and workstations are providing a significant market pull for the special techniques and faster types of memory devices such as SRAMs (static random access memory) necessary to operate at high speeds. If the HDTV market develops, it could provide an additional pull for leading-edge fast DRAMs. Matsushita's 9 Mb Video RAM, for example, has a serial access time of 20 ns, 1.5 times faster than current VRAM (video random access memory) technology.

HDTV could also become a significant new market for DRAMs. An HDTV consumer receiver might use as much as 32 million bits (Mb), equivalent to four million bytes (MB) of DRAM, to store the HDTV picture in memory. While some researchers believe that fairly high resolutions can be achieved without such large use of memory, others insist that there are significant advantages in having a full-frame memory, or 20 to 32 Mbs, depending on the standard, etc. Ultimately, the memory and digital signal processor demands will depend strongly on the particular standard chosen, but will also likely increase over time. Assuming a rapid growth of the HDTV market, then the use of DRAMs in HDTVs alone in 15 years could be five times the total 1987 world demand (by memory capacity, or bits) for all DRAM applications. Depending on the relative growth of the computer, telecommunications, and other markets, this may or may not represent a significant fraction of world

DRAM use at that time. Japanese firms are, however, already establishing major new DRAM production facilities with the expectation that their output will be used in advanced TVs. These scenarios for DRAMs and DSPs are subject to a number of qualifications and uncertainties. If the standard chosen for HDTV uses significantly less or more memory and digital signal processing, then the projections would need to be adjusted accordingly. If the development of the improved-definition television (IDTV) market, for example, substitutes for HDTV and prevents HDTV market growth, then the DRAM projections would be reduced by a factor of four because the typical IDTV is expected to use about a quarter of the memory used in an HDTV. If consumers instead move progressively upscale, buying IDTVs and extended- or enhanced-definition televisions (EDTVs) first and then move to HDTV, relegating their IDTVs/EDTVs to use as a second set as they did black-and-white sets when moving to color, then chip demands could be 25 to 50 percent greater than projected. If strong commercial markets for HDTV develop, as predicted by the Japanese Ministry of Posts and Telecommunications (MPT), then chip demand could be twice the projections above. Factoring in production and broadcasting equipment sales could increase these projections by 10 to 15 percent. Finally, some believe that progressively more sophisticated systems will be developed beyond HDTV, requiring even more memory and signal processing. Fifteen years ago the United States had more than 90 percent of the world market in DRAMs; today the United States makes less than 15 percent of the DRAMs purchased in world (merchant) markets. Texas Instruments, Micron Technologies, Motorola with technology licensed from Toshiba, and IBM (for internal consumption) are the only U.S. firms still producing DRAMs. The recent effort to form a consortium, U.S. Memories, might have improved somewhat the U.S. position. For a variety of reasons, however, it failed to attract sufficient support from U.S. firms even to be launched.

Receiver-compatible HDTV systems propose use of the standard 6-MHz NTSC signal, which would then be augmented with a second signal 3 to 6 MHz wide to provide the additional information for the higher-quality picture. The wider bandwidth of such HDTV systems may require GaAs (gallium arsenide) chips in the tuner due to their wider bandwidth capability and their ability to handle overloads.

GaAs and related materials are now used in a variety of applications, ranging from some leading-edge super-computers to the lasers in CD players and in fiber-optic systems. The use of GaAs remains limited, however, due to the difficulty of producing high-quality stock material and fabricating semiconductor devices from it. If HDTV provides a large market for GaAs devices, the additional production volume might help some of these difficulties to be overcome. Improved GaAs materials production and fabrication techniques could have spinoffs to a variety of markets.

The United States seriously lags behind Japan in a variety of GaAs and related materials, processing, and device technologies. Despite pioneering the develop-ment of many of these semiconductor devices, the United States today buys much of the unprocessed GaAs material from Japan as well as the semiconductor devices fabricated from it. AT&T invented the solid-state laser, but in some cases has purchased semiconductor lasers from Japan to drive its fiber cable.

Highly disciplined and cost-effective manufacturing is required for a large HDTV market to develop and for a firm to successfully compete within it. Technology will have to squeeze even more circuitry onto the same sliver of silicon to bring the cost of HDTVs to reasonable levels. Reductions of the total number of chips in this manner reduces costs—of components, of assembling and testing the HDTV, and of repairing defects, and by increasing reliability. Reducing the number of parts in color TVs was an important aspect of the competition between the U.S. and Japanese producers in the 1970s— and an aspect in which U.S. producers seriously lagged. The quest to reduce the number of chips in systems is responsible for the explosion in application-specific in-tegrated circuit (ASIC) production, which now accounts for about a quarter of all merchant integrated circuit production. (Merchant producers are those that sell on the open market and include all Japanese and most U.S. semiconductor producers. Captive producers are those that use the semiconductors they produce themselves and do not sell them outside the firm. The world's top three ASIC producers are Fujitsu, Toshiba, and NEC).

Efforts to reduce the number of chips needed can already be widely seen in ATV development. NEC, for example, reduced the number of chips in its IDTV from 1,800 to 30. An early prototype of the Japanese MUSE HDTV system had 40 printed circuit boards, each containing 200 chips for a total of 8,000 chips. In contrast, the generation of MUSE decoders unveiled in June, 1989, had less than 100 chips. Half of these were ASICs with 26 different designs. To minimize the burden on any one manufacturer in developing these numerous and complex ASIC designs, NHK divided the effort among six different manufacturers, and then distributed the designs among all the participants.

Manufacturers are also pushing the design of conventional memory chips. Matsushita recently unveiled an 8-Mb Video RAM designed specifically for application in HDTV, and intends to begin commercial sampling in 1990.

Eventually, nearly all of the required memory and DSP for an HDTV might be combined on a single chip. Increasing levels of chip integration will require a significant increase in current capabilities, and correspond to the expected leading edge of semiconductor technology for the next decade. The extent to which this drives the state of the art will depend on the relative size of the HDTV, computer, telecommunications, and other markets.

A number of important studies have documented the current U.S. lag behind Japan in a broad range of semiconductor process technologies:

- The Federal Interagency Task Force found the United States lagging behind Japan in 14 semiconductor process and product areas; the United States was ahead in just six categories and its lead was found to be slipping in five of these.

- The National Academy of Sciences found the Japanese leading in eight of 11 semiconductor process technologies that will be critical in the future.
- A 1988 study by the Department of Commerce found Japanese semiconductor plants had a five-year lead over the United States in the use of computer-integrated manufacturing techniques. These techniques have allowed the Japanese to reduce turnaround time by 42 percent, increase unit output by 50 percent, increase equipment uptime by 32 percent, and reduce direct labor requirements by 25 percent.

The Japanese have also rapidly improved their plant and equipment to take advantage of the more technically demanding but cost-saving large wafer technology. From a position of parity in 1984, they now use on average wafers that are nearly 35 percent larger in area than their American competitors. IBM, however, is pioneering very large, eight-inch, wafer technology.

The state of the art in magnetic and optical storage technologies for studio use is already being pushed by the large volume and high rate of information flow requirements for HDTV and, to a lesser extent, related HRS. Digital VCRs for studios will require much higher magnetic recording densities and information transfer rates through the use of improved magnetic materials, recording heads, and other techniques. Sony, for example, has developed a prototype studio VCR that has a recording speed of 1.2 Gbps—five times faster than previous VCRs were ever capable of. To similarly extend recording times on compact disks with video, the semiconductor lasers used will have to operate at higher frequencies than those used today, requiring advances in semiconductor lasers and reduction in production costs. Matsushita has recently succeeded in storing 2.6 Gb of video information on a single 12-inch optical disk. These recording technologies will have many applications in the computer industry.

Such spinoffs from consumer electronics have already been widely seen. Magnetic and optical (compact disk) storage technologies were both originally developed for the consumer electronics market, but are now used widely in the computer industry. In particular, compact disks are expected to have a profound impact on information handling generally.

Similarly, digital audio tape (DAT) machinery, originally developed for the consumer market, is expected to have a significant impact on the computer data storage market. DAT sales in the U.S. consumer market have been limited due to U.S.–Japan trade friction and issues of copyright protection; therefore, prices are expected to remain higher than if large volume sales had already been achieved. If approved, legislation pending in the Congress that requires copy-controlling devices in DAT machines may open up U.S. markets. DAT will be able to store about 1.3 Gb of data on a cassette the size of a credit card and about ⅜ inch thick and will have data rates of roughly 1.4 Mbps.

There are also spinoffs between technologies. The hard drives used in computers are made by coating a very thin, high-quality layer of magnetic material on a metal disk. The technology to do this, and even the processing equipment, originally came from semiconductor wafer fabrication. A key technology for large-area, flat-panel displays will similarly be putting extremely thin, high-precision coatings over very wide areas. Once developed, this could have an impact on the production of semiconductors, and on magnetic and optical storage. The converse is also true. Thin-film technologies developed for the semiconductor industry could initially have an impact on flat-panel production, although the impacts will likely decrease as the panel areas increase.

Precision motors like those used in HD-VCRs will also be used in computer tape and disk drives, robotics, and elsewhere. Today, Japan is the world's largest producer of precision motors due to this synergy of uses among electronic products.

The high-precision helical-scan drives for VCRs are now made primarily in Japan, although a few are made in Korea. Exabyte of Boulder, Colorado, purchases off-the-shelf 8mm camcorder-type tape drive mechanisms from Sony and uses them in a 2.3 Gb tape system (the highest storage capacity to date) for computer data storage; they are totally dependent on the Japanese source. With the continuing move to higher density storage systems, firms that produce computer tape storage systems, but do not have access to helical scanning (VCR-type) tape drives are unlikely to survive. The United States continues to hold a strong R&D and market position in some storage technologies, but has seriously lost ground in others. The United States has largely lost the floppy drive business, holding just 2 percent of world sales in 1987; but in the hard drive market, U.S. firms have fought back successfully and still hold a 60-percent or better share. 3M continues to be a major world-class producer of magnetic tape, but no U.S. firm produces the high-performance helical-scan recorder drives. Only one domestic firm, Recording Physics in California, has the capability to produce the very high-performance materials needed for read/write VCR heads.

The United States lags in many areas of optical storage research, and has little presence in the manufacture of optical storage devices. Over a dozen Japanese firms are developing or selling advanced rewritable optical disks and/or drives. The optical data storage device market is expected to grow from $400 million in the United States in 1988 to $7.3 billion by 1993.

Packaging/interconnect (P/I) is the set of technologies that connect all components—semiconductors, displays, storage, and communications—into functional systems. To connect a silicon chip to the outside world, the chip is mounted in a plastic or ceramic package that has tens to hundreds of metal leads. These packaged chips are mounted on printed circuit boards, which, in turn, are interconnected via standard multipin connectors on a motherboard or a backplane.

The costs of these connections increases rapidly at each level. Within the chip itself there are millions of tiny wires of aluminum connecting the transistors formed in the silicon. Despite their complexity, these connections typically cost just $0.0000001 each because they are all formed in a single step using a photomask. The cost of the connections between the chip and the package it is mounted on are roughly $0.01. The cost of connections between the package and the printed circuit board are roughly $0.10 each. Overall, the cost of packaging/ interconnecting and assembling the electronic components, together with testing the system, accounts for roughly 30 to 50 percent of the total for a complex electronic system. Most system reliability problems are due to interconnect failures, and P/I technologies are a principal barrier today to achieving higher system performance.

Today's printed circuit board technology, for example, etches individual circuit patterns in copper foil laminated to sheets of fiberglass-reinforced epoxy. Multiple layers of unique circuit patterns can be laminated together with more epoxy. Packaged semiconductors and other components are then mounted on the board and interconnected via copper-plated holes to specific circuit patterns on different layers of the board. These holes account for a large fraction of the total board area and limit wiring densities and component spacing, and thus slow attainable system speeds while increasing system size, weight, and costs. The mechanical drilling process used to form these holes limits further reductions in their size.

As an example, interconnecting the 200 or so chips of a supercomputer processor, each chip having 250 input/output leads or more, would require a board with 40 layers of circuitry. Advanced ICs may require 500 to 1,000 inches of interconnect wiring per square inch of board—two to three times the current practical limit. Designing and building such boards reliably is very difficult.

Small, high-density multichip modules are one means of improving P/I performance that is now gaining favor. IBM's 3090 mainframe, for example, combines many chips in a ceramic module with 44 layers of wiring. These modules are then mounted and interconnected via a printed circuit board with relatively few layers.

In the longer term, the Japanese Ministry of International Trade and Industry (MITI) and Key Technology Center flat-panel display consortium intends to use the lithography and thin-film technologies developed for large-area flat-panel displays to advance these printed circuit board densities through improved and lower-cost "chip-on-glass" technologies.

Chip-on-glass technology mounts "bare" integrated circuits directly on lithographically printed glass substrates. This has several important advantages. Fine-line lithographic printing can provide perhaps 10 times the wiring density attainable with the conventional copper-epoxy printed circuit boards described above. Increasing the wiring density also reduces the number of layers necessary. This reduces the space that must be allotted to the interconnections between layers. The savings are multiplicative. A conventional printed circuit board with 40 layers of copper-epoxy interconnect might be replace with a lithographically printed glass substrate with just two layers of interconnect. This provides substantial cost savings in both design and production.

Mounting the bare IC directly on the substrate bypasses several conventional packaging and interconnect steps with further corresponding cost savings and improvements in reliability. Together, the higher wiring density and use of bare chips can allow substantial increases in how close components are packed. This allows higher speeds and reduces system size and weight.

Chip-on-glass technologies are used in special high-performance cases today, but could be applied much more widely if large-area lithography tools and related technologies were available. These techniques would allow many glass substrates to be produced at once on a large sheet, rather than tediously one at a time.

The complexity and high speed of the chips used for HDTV will require the use of high-performance printed circuit boards. Complex multilayer printed circuit boards will be necessary and new materials may have to be developed to handle the high speeds at an affordable cost. Although these are all available today in high-end commercial and military markets, manufacturing in volume for the HDTV market might force rapid improvements in production technology and dramatically lower their price.

The United States seriously lags behind Japan in

many of these P/I and related assembly technologies. A 1988 National Academy of Sciences study found the majority of U.S. companies four to five years behind Japanese competitors in manufacturing process control and in factory automation for fabricating, assembling, and testing electronics products. The United States also lags behind Japan and Europe in the use of surface-mount technology for connecting the chip to the printed circuit board. This technology saves space, increases reliability and performance, and reduces assembly costs. Tape-automated bonding (TAB) technologies for packaging semiconductors, invented in the United States by GE but used more widely in Japan, offer significant increases in reliability at greatly reduced labor and cost. In addition, TAB significantly improves semiconductor performance.

Producing P/I equipment and materials for HDTV or, more generally, for the flat-panel display market may provide economies of scale to a firm as well. Shindo Denshi, the largest Japanese producer of TAB tape, currently gets half of its sales from supplying producers of LCD displays. Some Japanese companies, such as Toshiba and Matsushita, have also developed proprietary "outer lead bonders" for connecting wires to the display. This technology is not for sale and might make it more difficult for U.S. firms to enter the market.

Many P/I and related technologies—assembly, test, surface-mount, tape-automated bonding—have been pushed the hardest by the consumer-electronics market. The Sony Watchman television, for example, uses higher performance TAB than the NEC SX-2 supercomputer. The consumer electronics market demands high reliability, small size, and low cost, but at the same time provides very large volume markets that allow even expensive, yet innovative, technologies to pay for themselves through long-term productivity improvements as experience is gained.

Because of these characteristics, consumer electronics often pushes the state of the art in manufacturing technologies harder than lower-volume but higher-profit markets—especially for assembling components or systems. If the HDTV market develops, it may similarly provide manufacturers a testing ground for developing new assembly technologies with the volume needed to pay for themselves, as well as gaining valuable experience in assembly of sophisticated electronic systems that can be transferred to many other products.

*Editor's note:* This chapter was excerpted from "The Big Picture: HDTV and High Resolution Systems," a background paper published by the Congress of the United States, Office of Technology Assessment, June, 1990.

# 11

## Advanced Television:
## Opportunity for U.S. Industry

*Dr. D. Joseph Donahue*
*Senior V.P., Technology and Business Development,*
*Thomson Consumer Electronics*

A s we begin the 1990s, a debate continues in the U.S. over how best to respond to the competitive challenges posed by Japan and the European Community, particularly in high-technology industries. Since the late '80s, high-definition television (HDTV) has been the focal point—some would say the flash point—for much of that emotion-charged discussion, which has attracted considerable industry, government, and media attention.

Contrary to the conventional wisdom, the United States today is well on its way toward developing advanced television (ATV) systems tailored to its own unique requirements. The much-publicized technological lead enjoyed by the Japanese and Europeans, while substantial in terms of their own markets, has little relevance to the search for an ATV standard in this country. The fact is that the United States has embarked on a policy course that promises to result in the introduction of viable ATV systems by the mid-1990s.

Much has been written about the potential size of the ATV market in the U.S. While these estimates vary widely, most observers agree that in TV receiver sales alone the American market will amount to tens of billions of dollars over the next two decades. In my view, most of the resulting economic benefits will accrue to the American economy, assuming that the affected industries work together toward a common goal and that the federal government provides a hospitable policy environment.

Among the widely held myths surrounding the consumer electronics industry today are that color-TV production takes place off shore, and that research and development in the U.S. has virtually disappeared. Policy makers and journalists routinely refer to the consumer-electronics sector as an example of an industry that has been "lost" or has gone the way of the brontosaurus.

As is so often the case, the conventional wisdom is unreliable. Consumer electronics in the United States is a $45-billion business (at retail), and color television remains its flagship product. While it is true that a number of U.S.-owned companies have left the color TV business in recent years, the fact is that virtually every world-class manufacturer has established at least one or more production facilities in the U.S.

American consumers purchased approximately 22 million color TV receivers in 1989, with some three dozen brands accounting for 95 percent of that total. It may come as a surprise that approximately two out of every three sets purchased by U.S. consumers are produced in this country by American workers. And in the 25-inch-and-larger category, more than nine out of every 10 color sets sold in the U.S. are produced in the U.S. as well.

My own company, Thomson Consumer Electronics, Inc., whose RCA and GE brands consistently account for the largest market share of color TVs, produces color TV receivers—as well as picture tubes, picture-tube glass, PC boards, and plastic and wood cabinets—at a half dozen U.S. manufacturing plants.

It is equally false to assume that major consumer electronics manufacturers rely on overseas facilities for their R&D. Global players such as Thomson and Philips—whose brands account for nearly 40 percent of total color TV sales in America—are deeply involved in R&D in general and ATV in particular at their U.S.-based research laboratories. While Thomson and Philips continue to explore new television technologies at their European R&D centers, both firms have given their U.S.-based R&D facilities full responsibility for developing advanced television systems for the American market.

Today there appears to be a growing understanding of the extent to which consumer electronics is a global industry, and that it borders on fantasy to suggest that a U.S.-owned consumer-electronics industry can somehow be resurrected. The fact is that national borders have become largely irrelevant not only in consumer electronics, but in much of the electronics industry as a whole.

On this question of ownership by nationality, Robert B. Reich argues in the January-February 1990 issue of the *Harvard Business Review* that the Thomsons and Philipses of the world contribute at least as much, and probably more, to the strength and competitiveness of the American economy than many of their U.S.-owned counterparts. Dr. Reich poses the issue this way:

> So who is us? The answer is, the American work force, the American people, but not particularly the American corporation. The implications of this new answer are clear: If we hope to revitalize the competitive performance of the United States economy, we must invest in people, not in nationally defined corporations. We must open our borders to investors from around the world rather than favoring companies that may simply fly the U.S. flag....

Dr. Reich's views are reinforced by Todd Hixon and Ranch Kimball of the Boston Consulting Group, who conclude that:

> The real payoff from local operations for foreign-owned companies, then, comes in the form of fully integrated business operations—when product design, process design, manufacturing, and vendor management are co-located and tightly integrated in-country and the operation is set up to do business in the global market....[1]

This view is reflected in the policies of the Bush Administration. A letter to the Chairman of the House Subcommittee on Science, Research and Technology from the Commerce Department's General Counsel takes the position that "any legislation which contains provisions requiring the identification of 'foreign' firms and the discriminatory treatment of such firms would not be in accord with the President's program."

Ownership issues should now be placed behind us. All parties committed to strong R&D and manufacturing investments in the United States should be fully eligible for any government support.

Instead of agonizing over ownership by nationality, our focus should be on how to create the maximum number of American jobs in advanced television R&D and manufacturing, and how to increase, to the maximum feasible extent, the amount of U.S. content in advanced television products.

During the 35-year span from the immediate postwar period to approximately 1980, the consumer-electronics industry experienced widespread consumer acceptance of black-and-white television and then color TV. It is fair to say, however, that despite sales growth and improvements in programming, program delivery, and receiver technology, there was relatively little diversity in product features or services during that period.

The decade of the '80s may have marked a turning point for the television industry. With such innovations as cable, direct-broadcast satellite (DBS), and prerecorded video, we have witnessed nothing less than a revolution in the ways in which programming is delivered to the home. Similarly, diversification of services has increased, including a higher number of channels and the introduction of multichannel sound (MTS).

As services have increased, so have the number and variety of the TV sets' capabilities, including larger screen sizes (in both projection and direct-view), remote control, on-screen displays, enhanced picture quality, and digital special effects. When we add videocassette recorders (VCRs), which are now owned by at least two of every three U.S. households, and camcorders, the fastest-growing video hardware product, we realize the extent to which we have experienced a home video revolution.

The industry's dynamism and diversification seem

likely to continue, given the consumer's enormous appetite for new products and programming, the rapid technological progress currently underway, and the relative decline in government regulation of the delivery media.

In this kind of environment, which clearly seems conducive to growth and innovation, it is reasonable to expect an acceleration of this trend, particularly in the areas of fiber optics, improved video and data services, filmless electronic photography (via the TV set), and advanced displays, not to mention enhanced-definition television (EDTV) and HDTV.

The critical question facing industry executives and government policy makers is not whether these advances in TV technology are coming. They are. The more important question is how successfully we will manage their inevitable introduction.

A fundamental requirement is that we agree on a minimum number of basic standards in order for all the affected industries to move forward with confidence. When the involved parties agree upon a single set of standards and provide strong program support—as they did in the cases of the compact disk (CD) and stereo TV technologies of the 1980s—the product has an excellent chance of flourishing. On the other hand, when we fail to reach a consensus, implementation often falters; witness the cases of multiple videodisc formats and AM stereo.

There is considerable skepticism about whether advanced television will offer enough tangible and perceived advantages to win widespread consumer acceptance, especially in view of the anticipated premium prices of first-generation receivers.

We can make the next generation of television attractive to large numbers of consumers only if receiver manufacturers work closely and cooperatively with the various delivery industries to solve the "chicken-and-egg" problem.

As a young engineer in the early '50s, I learned firsthand how vital this cooperation was in launching color television in the United States. Inadequate industry support of color TV delayed consumer acceptance by nearly a decade; it was not until 1964 that our industry enjoyed its first million-unit sales year for color receivers.

A more recent and instructive new product introduction took place in the mid-'80s when broadcasters and manufacturers introduced stereo TV in the U.S. During 1989, just five years after its introduction, more than six million stereo-equipped color sets were purchased by American consumers. Not only does stereo TV offer an excellent example of broadcasters and manufacturers overcoming the "chicken-and-egg" problem, it represents a classic example of successful government-industry cooperation. By agreeing to a unified industry recommendation of an engineering standard for stereo television, the Federal Communications Commission (FCC) was instrumental in helping to get this exciting new technology off the ground.

For their part, consumers will accept technological advances, not for technology's sake but because these advances offer a combination of improved hardware and software. And for that to occur, consumer-electronics manufacturers and the various delivery media must reach a consensus.

If all key industries work together, we can move into the age of advanced television in a positive, evolutionary way. Conversely, confusing or inadequate presentation to the consumer could mean squandering human and financial resources, thereby delaying the introduction of these new technologies.

We stand on the threshold of an exciting new era—the age of advanced television. The challenge facing us is to present this new technology to consumers clearly and confidently. This will ensure that consumers share fully in the benefits of innovation and increased competition.

## Endnote

1. Copyright ©1990 by the President and Fellows of Harvard College; all rights reserved.

# 12

# Consumer Electronics, HDTV, and the Competitiveness of the U.S. Economy

*Advanced Television Committee,*
*Electronic Industries Association*

*T*here is widespread interest in HDTV for three reasons.

First, the HDTV market is projected to be large, and the potential effects of HDTV on national production, employment, and trade performance are estimated to be substantial.

Second, HDTV is viewed by some as a way for the U.S. to re-enter television and VCR markets—which U.S-owned firms, with few exceptions, have virtually abandoned. According to some observers, the costs of this abandonment of a major part of the consumer-electronics market have been lost production, sales, employment, and a dramatic deterioration in the U.S. trade balance in electronics. These costs, however, have been greatly exaggerated.

Third, there is growing policy concern about U.S. participation in HDTV due to the technological spillovers that HDTV production may generate in a variety of related activities, including the development of new semiconductor componentry and new video display technology. These spillovers could affect the American competitive position in a variety of important industries, including computers and advanced telecommunications equipment. Spillovers with application to defense are also considered likely.

It is claimed that there have been additional costs in linked industries such as semiconductors. The dramatic drop in the U.S. share of world semiconductor sales is, in part, the result of the increasing content of semiconductors in consumer electronics (televisions, radios, disk players, electronic games, etc.), a business that companies based in Japan dominate. At least 35 percent of Japanese consumer-electronics production has been sold to the U.S. in consumer-electronics products as have been sold to the U.S. directly. But recapturing semiconductor market shares through a revived U.S. consumer-electronics industry will not be easy because that industry will continue to be competitive and thus will behave more or less as it has in the past—companies will source their electronic components on the basis of price, performance, and quality.

To many the question of whether the U.S. will be able to occupy a competitive position in the emerging HDTV market and related technologies has become symbolic of the broader question of whether the U.S will be able to regain its national competitive strength. Increasingly, the competitiveness of the nation has become associated with its ability to emerge a winner in the HDTV market. This symbolism is seriously misleading.

A fundamental premise of the Electronic Industries Association's Advanced Television Committee is that competitiveness is primarily an economy-wide issue and is logically distinct from the competitive position of the nation's producers in a particular industry or activity. A corollary is that the most effective policies to improve national competitiveness must address broad-based problems, such as the low rates of national saving and investment, the high cost of capital, an inadequately educated and skilled workforce, and insufficient public support for generic or middle-ground R&D, all of which adversely affect private sectors across the industrial spectrum.

An aggregate perspective, however, overlooks the reality that certain industries or activities may contribute more than others to national competitiveness over the long run. And there is evidence to suggest that the electronics sector, broadly defined to include the semiconductor industry, the telecommunications industry, the computer industry, and at least segments of the consumer-electronics industry, falls into this category. Many policy makers and industry participants believe that the long-term competitive health of many parts of the U.S. electronics sector will be adversely affected by a U.S. competitive failure in HDTV.

Even if U.S. policy makers and industry representatives are persuaded that the U.S. competitive position in the emerging HDTV area is important to the nation's long-run competitiveness, many unresolved policy issues remain. Two such issues are of paramount importance. The first arises because of the particular ownership configuration of television set producers in the U.S. Currently, most producers are foreign-owned, but many of the foreign-owned facilities operating here have broad-based activities, ranging from R&D to distribution. Many conduct extensive R&D activity in the United States, an activity that should be encouraged by government policy.

This raises a fundamental question that may be addressed before policy decisions are made, to wit: If the objective of policy is to foster U.S. participation in HDTV, will participation by foreign-owned firms operating from U.S. locations support this objective? In other words, does U.S. participation mean participation by domestically owned firms regardless of where they locate their production, employment, and research facilities, or does it mean participation by foreign-owned firms operating the U.S.—or perhaps does it mean a combination of both? The HDTV issue reveals the increasingly global nature of many high-technology industries and the difficulties of making public policies to foster national participation. To limit such policies to domestically owned companies is likely to delay the development and introduction of HDTV technology in the U.S. market and to discourage foreign-owned producers from expanding their production and R&D operations in the U.S.

A second basic policy question is whether policies

specifically targeted to HDTV are required to foster U.S. participation, however defined. Perhaps a combination of broad-based procompetitiveness policies, such as a change in the monetary and fiscal mix with lower interest rates, an R&D tax credit, further relaxation of antitrust limitations on joint R&D activity, and continued efforts to insure fair competition in U.S. and international markets, is all that is needed. Certainly, without a change in at least some of these broader policy areas, it is unlikely that all the beneficial effects of U.S. participation in HDTV on U.S. competitiveness will be realized.

Finally, even if special policies to foster U.S. participation are required, what form should they take? What are the appropriate roles of standard setting, R&D consortia, Defense Department spending, and other policies?

During the last several years, the dramatic and sustained deterioration in the U.S. trade deficit has created growing concern over American competitiveness. Indeed, competitiveness, a concept that did not even exist in national policy discussions five years ago, has become a buzzword. Business, labor, education, and government leaders speak of the competitiveness challenge confronting the United States and offer a potpourri of sometimes conflicting policy solutions. Initiatives in such diverse areas as trade legislation, education reform, and taxes are defended or criticized on the bias of their effects on U.S. competitiveness.

Like most buzzwords, competitiveness has symbolic significance. It draws national attention to the undeniable fact that the position of the United States in the world economy is weakening. America can no longer rest comfortably in the belief that it will continue to be the leading economic power in the world. Although still the largest and one of the richest economies, the United States has lost ground compared to many countries with which it competes in world trade. To some extent, of course, this was inevitable. As the other developed countries rebuilt from war destruction and as many developing countries introduced ambitious development programs—helped by American funds and American technology—some catching up was inevitable. But the pace and extent of the catch-up—or to put it differently, the pace and extent of the relative decline in the U.S. position—were not inevitable. And significantly for the future, there is not inevitability of a continued decline in the U.S. position.

If competitiveness is to have more than symbolic significance, if it is to become a reliable guide for policy, it must be properly defined. The EIA/ATV Committee defines competitiveness, as it was defined in the Report of the President's Commission on Industrial Competitiveness, as "the degree to which a nation, under free and fair market conditions, produces goods and services that meet the test of international markets while simultaneously maintaining and expanding the real incomes of its citizens."[1] There are four important points about this definition that should be emphasized.

First, competitiveness implies an ability to compete in international markets, with balanced trade over the long run, without an associated decline in real wages, and without a continued decline in the value of the dollar that would cause falling real wages over time. Competitiveness is not simply the ability to sell abroad or to maintain a sustainable trade position at some exchange rate. The very poorest nations often boost exports just by devaluing their currencies. The consequences, however, are sharp declines in relative wages and relative standards of living, declines that are at odds with national competitiveness as defined here.

Second, competitiveness implies an ability to compete in free and open markets. In this sense, neither a worsening of U.S. trade performance occasioned by unfair trade measures abroad nor an improvement occasioned by protectionist trade measures at home is a sign of a real change in U.S. competitiveness.

Third, competitiveness is a concept related to economy-wide performance. Policy discussions sometimes focus on the competitive position of particular U.S. industries, such as the U.S. semiconductor or consumer-electronics industry. These performance criteria are taken as indicative of U.S. competitiveness broadly understood. While certain industries or activities may be especially important to national competitiveness, competitiveness is best understood as an economy-wide concept rather than as a sectoral or industrial one. And the most effective policies for improving national competitiveness are broad-based policies aimed at strengthening national productivity and technological performance—the two basic determinants of national competitiveness.

Finally, competitiveness is a national concept. It is concerned with the relative economic performance of nations, not companies. The focus on national economic performance means that indicators of national competitiveness, rather than indicators of the competitiveness of individual firms whose headquarters are in one nation but whose production and distribution facilities are internationally located, are the relevant subject for policy concern. According to this perspective, domestic firms can adopt production, investment, location, and sourcing strategies that weaken national competitiveness, while foreign-owned firms operating in the U.S. can adopt comparable strategies that actually strengthen the competitiveness of the United States.

Just as the choices of American multinationals have contributed to the growing competitiveness of the East Asian newly industrializing countries, so the choices of foreign multinationals operating in the United States may contribute to the restoration of American competitiveness. In short, there is no simple or necessary relationship between the ownership of firms operating in a nation and its competitiveness as defined here.

There are three important forms of linkage between the consumer-electronics industry and the rest of the electronics complex. They are upstream effects, downstream effects, and manufacturing effects.

Upstream effects derive mainly from the role of consumer-electronics production as a source of demand for inputs, and, in particular, for semiconductor components. The consumer-electronics industry in the United States first contracted and then shifted from domestic to predominantly foreign ownership. The ability and interest of U.S.-based semiconductor firms to service markets for consumer-related semiconductors virtually disappeared.[2] By the mid-1980s, only 6 percent of semiconductor production in the U.S. went to consumer applications, whereas in Japan, 40 percent did. In dollar terms, this meant that Japan was producing 7.2 billion consumer chips in 1987 while the U.S. produced only 0.9 billion. The corresponding figure for Europe was around four billion.[3]

There is an honest dispute about how this occurred. Some U.S. firms claim that foreign-owned consumer electronics firms had preferential supply arrangements that excluded them from the market. The more vertically integrated foreign electronics firms often sourced their semiconductors from their internal semiconductor divisions. In all the major industrialized regions, there is a preference for working with regional suppliers of components whenever possible. The Japanese consumer-electronics industry, as represented by the EIAJ, claims that U.S. firms were unable to produce the necessary products, or to deliver them on time, or to match the quality/reliability of other (particularly Japanese) producers. The U.S. semiconductor firms accuse the Japanese of preferentially sourcing from Japanese semiconductor producers. Both of these claims may be true. Of key importance for many U.S. firms was the fact that the consumer chip business was less profitable than business for industrial or defense applications, because it involved standard devices in which markets were highly competitive.

Another key factor was the increased product design activity that built up in Japan. Product design in a foreign country makes it extremely difficult for U.S. semiconductor firms to compete. It is estimated that around 12 percent of the semiconductors produced in Japan are used in VCRs. The end result of these two factors was major abandonment of consumer chip production in the U.S. This is particularly true in such specialty devices as charge-coupled devices (CCDs) and liquid crystal displays (LCDs).

Downstream effects refer to the impact of consumer electronics on industries downstream from the semiconductor industry. The VCR volume base led to video cameras. The camera base contributed to commercially priced CCD chips. The VCR, camera, and CCD base led to camcorders, still video photography, video printers, and new video printer film.

In a similar manner, Japanese strengths in LCDs for watches and calculators helped to give them an edge in the emerging markets for laptop computers and personal TVs.

Manufacturing effects involve the loss of strength in generic manufacturing skills and technologies associated with the reduced role of U.S.-owned firms in the consumer-electronics industry. While a number of U.S. firms were able to match their international competitors in the adoption of advanced manufacturing techniques, such as automated insertion and surface-mount technologies, the majority failed to do this rapidly enough to meet the competition. These technologies are important not just for consumer-electronics but for many other kinds of high-volume production. The decline of the U.S. consumer-electronics industry, therefore, meant a narrowing of the manufacturing skill base of the U.S. economy.

There are reasons to believe that upstream, downstream, and manufacturing effects will be even greater in the next two decades than they were in the past.[4] HDTV circuitry will be much more complex than NTSC circuitry. HDTV circuitry needs could contribute to advancing technology in some important areas, such as digital signal and image processing and parallel processing. HDTV receivers will require larger and better video frame storage devices than NTSC receivers. In addition, competition in the HDTV business will create sizeable incentives for the development of large displays, and particularly for the development of flat-panel displays—e.g., LCDs, and semiconductor-based projection systems.

The downstream spillover effects of HDTV technology could be significant in the computer, defense electronics, and telecommunications industries. The problems of image and digital signal processing that have to be solved for HDTV receivers also have to be solved for fast displays of color images on advanced computer workstations. The production of large, high-resolution displays for HDTV equipment will allow some firms to produce cheaper and more competitive displays for computers and workstations.

There is an important mutually reinforcing relationship between advances in HDTV and network (telecommunications) technology. The networking of advanced computer workstations creates network architecture design problems similar to those posed by the use of HDTV receivers as interactive terminals. Interactive video and interactive 3D color CAD/CAM are both more demanding than existing interactive character and graphics networking.[5] If you can solve one problem, then you have more or less contributed to the solution of the other. The unanswered question in this equation is how much demand there will be for "interactive" (two-way) as opposed to "passive" (one-way) television.

More important than the technological linkages between HDTV and telecommunications are the likely linkages between the two that arise with the building of a new national telecommunications infrastructure based on optical fibers. HDTV signals will be delivered to the home before the fiber-optic network is universally operational. Nevertheless, the sooner HDTV broadcasts and other home deliveries begin, the sooner there will be demand for transmitting HDTV signals via optical fiber (because of the greater fiber-optic bandwidth and the opportunity for reducing transmission noise with broadband digital signals). By the same token, the faster high-quality fiber-optic delivery to the home is in place, the easier it will be to convince consumers to make the switch from current NTSC, or interim products, to HDTV.

The greater U.S. participation in HDTV consumer markets is, therefore, the greater the upstream, downstream, and manufacturing benefits for the rest of the U.S. economy will be. Thus, policy measures should be aimed at maximizing U.S. participation. Because U.S.-based foreign-owned firms already possess such an important stake in this country's R&D and manufacturing of consumer electronics, they should be included in efforts to promote the HDTV industry.[6]

Factory sales of consumer-electronics products were around $31.6 billion in 1989. Of this total, 42 percent were derived from sales of TVs, VCRs, and camcorders (see Table 12-1). Over 20 million color TV sets, 11.6 million VCRs, and 1.6 million camcorders were sold in the United States in 1987.

**TABLE 12-1**
**Factory Sales of Consumer Electronics Products**
**in the United States, 1977-1988**
(in millions of dollars, including imports)

| Year | Mono TVs | Color TVs | Proj. TVs | VCRs | Video Discs | Audio Systems | Audio Comp. |
|------|------|------|------|------|------|------|------|
| 1977 | 530 | 3289 |  | 180 |  | 606 | 1275 |
| 1978 | 549 | 3674 |  | 326 |  | 748 | 1143 |
| 1979 | 561 | 3685 |  | 389 |  | 748 | 1178 |
| 1980 | 588 | 4210 |  | 621 |  | 809 | 1424 |
| 1981 | 505 | 4349 | 287 | 1127 | 55 | 720 | 1363 |
| 1982 | 507 | 4253 | 236 | 1303 | 54 | 573 | 1181 |
| 1983 | 465 | 5002 | 268 | 2162 | 81 | 630 | 1268 |
| 1984 | 419 | 5538 | 385 | 3585 | 45 | 976 | 913 |
| 1985 | 309 | 5562 | 488 | 4738 | 23 | 1372 | 1132 |
| 1986 | 328 | 6024 | 529 | 5258 | 26 | 1370 | 1358 |
| 1987 | 287 | 6271 | 527 | 5093 | 30 | 1048 | 1715 |
| 1988 | 236 | 6277 | 529 | 4820 | 40 | 1225 | 1854 |
| 1989 | 156 | 6530 | 478 | 4632 | 59 | 1217 | 1871 |

*Source:* EIA, 1987 Electronic Market Data Book, p.6; EIA, Consumer Electronics U.S. Sales, January 1989.

Several of the largest U.S.-owned firms in consumer electronics were purchased by foreign firms. RCA was purchased first by General Electric in 1985, and then sold to Thomson of France in 1987. Philips of the Netherlands purchased Magnavox in 1975, and Philco and

Sylvania in 1981. Zenith remains the only major U.S.-owned producer of TVs. In 1987, Thomson, Zenith, and Philips were the "big three" firms and accounted for about half of all color TVs sold in the United States. The rest of the market was divided among mostly Japanese and Korean producers. Both U.S.-owned and foreign-owned firms contribute to U.S. competitiveness in a variety of ways. There is significant variance in the degree to which each firm locates its research and development, manufacturing, and component production/sourcing in the United States (see Table 12-2).

Despite the increased participation of foreign-owned firms in the United States, the color TV part of the consumer-electronics market still retains a significant proportion of local content. The domestic manufacturing content of the average color TV made in the United States in 1987 is estimated to be around 70 percent,[7] and there has been a reversal in the last year or so of the downward trend in domestic content thanks to increased use of picture tubes manufactured in the United States. A number of foreign-owned tube manufacturing facilities came on line, a development that owes much to the decline in the value of the dollar relative to Asian and European currencies. The foreign content of TV sets is primarily in the electronic circuitry, but there is also some foreign content in the form of license and royalty payments for tube and chassis technology.

Competitiveness is primarily an economy-wide issue. There is a danger connected with equating the competitiveness of a nation with that of a single industry. While a single industry may be symbolic of general, national problems of competitiveness, certain policies designed to promote the revival of such symbolic industries may be prejudicial to the solution of the wider problem of competitiveness. For this reason, we recommend a judicious combination of economy-wide measures and industry-specific efforts. We recommend that economy-wide policies should focus on increasing investment levels in physical plant, human, and knowledge capital. In the case of measures specific to consumer electronics and HDTV, we recommend that only those that are likely to result in positive spin-offs for

# TABLE 12-2
## U.S. Set and Tube Production (1988 Data)

| COMPANY | LOCATION | NO. EMPLOYEES | PLANT TYPE | ANNUAL PRODUCTION | EXPORT/WHERE? |
|---|---|---|---|---|---|
| Bang & Olufsen (Joint Venture with Hitachi) | Compton, CA | (See Hitachi) | Assembly | Not Available | yes/Canada |
| * Goldstar | Huntsville, AL | 400 | Total Production | 1 million | yes/Taiwan |
| Harvey Industries | Athens, TX | 900 | Cabinet Assembly/ TV Assembly | 600,000 capacity | yes/Mexico, Far East |
| * Hitachi | Anaheim, CA | 900 | Total Production for 24", 27", 31" | over 360,000 | no |
| JVC | Elmwood Park, NJ | 100 | Total | 480,000 | yes/Canada |
| Matsushita | Franklin Park, IL | 800 | Assembly | 1 million capacity | yes/Japan |
| American Kotobuki (Subsidiary of Matsushita) | Vancouver, WA | 200 | VCR/TV Assembly | Not Available | Not Available |
| Matsushita (Joint Venture w/ Philips) | Troy, OH | 1-200 at opening; eventually 400 | Tubes | 1 million | Not Available |

*Continued*

| COMPANY | LOCATION | NO. EMPLOYEES | PLANT TYPE | ANNUAL PRODUCTION | EXPORT/WHERE? |
|---|---|---|---|---|---|
| Mitsubishi | Santa Ana, CA | 550 | Final Assembly | 400,000 | no |
| Mitsubishi | Braselton, GA | 300 | Total/Full | 285,000 | no |
| NEC | McDonnough, GA | 400 | Final Assembly | 240,000 | no |
| O-I/NEG TV Products (Joint Venture of Owens-Illinois & Nippon Electric Glass) | Columbus, OH | 800 | Tubes (capabilities up to 45") | Not Available | no |
| O-I/NEG TV Products | Pittston, PA | 750 | Tubes (capabilities up to 45") | Not Available | no |
| O-I/NEG TV Products | Perrysburg, OH | 75 | Component Glass for TV (Solder Glass) | Not Available | no |
| Orion | Princeton, IN | 250 | Assembly | Not Available | yes |
| Philips | Arden, NC | 400-500 | Parts (Plastic Cabinets and Accessories) | Not Applicable | yes/Mexico, Canada |
| Philips | Greenville, TN | 3,200 | Assembly/Full Manufacturing | Over 2 million | yes/Canada, Taiwan & Mexico |

*Continued*

| COMPANY | LOCATION | NO. EMPLOYEES | PLANT TYPE | ANNUAL PRODUCTION | EXPORT/WHERE? |
|---|---|---|---|---|---|
| Philips | Jefferson City, TN | 1,000 | Parts (Wood Cabinets) | 6-700,000 | no |
| Philips | Ottawa, OH & Emporium, PA | 2,300 | Tubes | 3 million | no |
| Samsung | Saddlebrook, NJ | 250 | Production for 13"-26" TVs | 1 million capacity | yes/Canada |
| Sanyo | Forrest City, AR | 400 | Assembly | 1 million capacity | no |
| Sharp | Memphis, TN | 770 | Assembly | 1.1 million | no |
| Sony | San Diego, CA | 1,500 | Full Manufacturing of Color TVs & Tubes | 1 million | yes/Canada, South America, Central America Taiwan & Japan |
| Tatung | Long Beach, CA | 130 | Assembly | 17,500 | yes/Canada, Mexico & Taiwan |
| Thomson | Bloomington, IN | 1,766 | Full Manufacturing/ Assembly | over 3 million | yes/Latin & South America |
| Thomson | Indianapolis, IN | 1,604 | Components Manufacturing (Printed Boards & Cabinet Production) | Not Applicable | no |

*Continued*

| COMPANY | LOCATION | NO. EMPLOYEES | PLANT TYPE | ANNUAL PRODUCTION | EXPORT/WHERE? |
|---|---|---|---|---|---|
| Thomson | Mocksville, NC | 626 | Cabinet Production | Not Applicable | no |
| Thomson | Marion, IN | 1,982 | Tubes | Not Available | yes/Europe |
| Thomson | Circleville, OH | 700 | Glass for Picture Tubes | Not Available | no |
| Thomson | Scranton, PA | 1,242 | Tubes | Not Available | yes/Europe |
| Toshiba | Lebanon, TN | 600 (300 add'l planned for '89) | Assembly | 900,000 (1.3 million planned for '89) | yes/Canada, South America, Japan & Taiwan |
| Toshiba | Horseheads, NY | 1,000 (500 add'l planned for '89) | Tubes | 1 million (1.5-2 million planned for '89) | no/(probably will export in '89) |
| Toshiba | Springfield, MO | 2,000-2,500 | Full Manufacturing/Final Assembly | Not Available | yes/Canada |
| Zenith | Melrose Park, IL | 2,500-3,000 | Tubes | Not Available | yes/Far East |

*Source:* Electronic Industries Association, HDTV Information Center

* indicates unverified data.

other industries should be the focus of public policies.

The development and commercialization of HDTV in the United States is an opportunity for the strengthening of U.S. competitiveness in the electronics complex and manufacturing more generally. HDTV is not the answer to all of America's problems in competitiveness, but it can contribute to their solution. Given the major presence of foreign-owned firms in the United States, there is an opportunity to build U.S. competitiveness with the help of those firms. Two main types of public policies are required to promote the HDTV industry in the United States: timely adoption of HDTV standards and assistance in the formation of R&D consortia to develop indigenous HDTV technologies.

## Endnotes

1. President's Commission on Industrial Competitiveness, *Global Competition: The New Reality* (Washington, D.C.: USGPO, December 1984); Council on Competitiveness, *America's Competitive Crisis: Confronting the New Reality* (Washington, D.C.: March 1987).

2. It should be noted that the decision of U.S. semiconductor firms to stop building chips for consumer electronics at the beginning of the massive growth in imports of consumer products occurred much earlier than the acquisition of major U.S. consumer firms by foreign firms.

3. Statement by Jeffrey A. Hart submitted to the subcommittee on Telecommunications of the House Committee on Energy and Commerce, September 7, 1988, p.8.

4. Competitive issues are also raised by the efficient use of the broadcast spectrum. To the extent the Federal Communications Commission allocates spectrum efficiently, new opportunities for growth in the U.S.-based land mobile and cellular telephone industry will be created. An important issue in the development of HDTV is the bandwidth proposed for each channel and the channel spacing. The EIA believes that the FCC should consider spectrum efficiency and HDTV terrestrial signal quality required to be competitive with other delivery media when it adopts the terrestrial HDTV transmission standard.

5. Workstation firms are now introducing NTSC video image

processing in the high end of their product lines.

6. Joseph Donahue, of Thomson Consumer Electronics in Indianapolis, estimates that the annual R&D expenditures of Thomson, Zenith, and Philips in the United States are around $150 million per year.

7. This estimate is based on figures for U.S. content of color TV receivers manufactured by Thomson. In Thomson's case, the U.S. content was 74 percent for 20-inch direct-view receivers, 77 percent for 26-inch direct-view receivers, and will be 82 percent for 31-inch direct-view receivers. U.S. contents increase in sets with the larger pictures because the tubes are mostly manufactured in the United States and they become a larger proportion of the total manufacturing cost of the larger sets. The domestic content of receivers produced by other firms may be somewhat lower than Thomson, but 70 percent is a reasonable estimate given the practices of other firms.

*Editor's note:* This chapter was excerpted from "Consumer Electronics, HDTV and Competitiveness of the U.S. Economy," a report by the EIA's Advanced Television Committee, submitted to the House Telecommunications and Finance Subcommittee, February, 1989.

# 13

# HDTV: America's Last Chance
# to Remain Competitive?

*J.J. Barry*
*International President,*
*International Brotherhood of Electrical Workers*

*A*merica has always been a nation of risk takers,
innovators, pioneers, and give-no-quarter competitors. It is this spirit that has lifted the country to greatness and made it the world's number-one
economic power.

Today, the U.S. is in grave danger of losing this
status. However, one simple decision by the government would help it regain its dominant position.

The issue is high-definition television (HDTV). The
choice facing policy makers is whether or not to assist
American HDTV manufacturers.

HDTV is the next revolution in consumer electronics. It provides pictures with twice the clarity of regular
television, and compact disk–quality sound. Its impact
will be comparable to the introduction of color television and of the videocassette recorder.

In the immediate future, the Bush Administration's
and Congress's HDTV decisions will determine whether
the American "can-do" spirit is still alive—or whether the
U.S. has become a nation of quitters.

Unfortunately, the White House appears to be
choosing the latter course, sacrificing the country's
national security and its economic future on the altar of
laissez-faire, free-market ideology. The White House cut

the Defense Advanced Research Projects Agency's (DARPA's) HDTV budget by two-thirds. And to make matters worse, the Pentagon ousted the agency's director, Dr. Craig I. Fields, because of his advocacy of HDTV. On the surface, the stakes may not seem so high. But in fact they are, for the following reasons:

1. *HDTV is critical to the nation's defense and to such important fields as aviation and medicine.* The technological breakthrough of HDTV has enormous and valuable applications in military electronics, surveillance, simulations, graphics, robotics, air-traffic control, high-tech surgical and diagnostic procedures, and other areas. That's why Congress appropriated $30 million this fiscal year (later cut to $10 million by the Administration) for DARPA to grant to companies developing high-resolution video display screens or signal processors.

   Without American HDTV manufacturing—which effectively means without an American commercial electronics industry—a major portion of U.S. defense becomes dependent on foreign countries. Could the U.S. have even fought World War II, much less won it, if it depended at the time on Japan and Germany for much of its military technology?

2. *HDTV is critical to the vital semiconductor industry.* Semiconductors are fundamental to HDTV technology, forming by some estimates about 20 percent of an HDTV set. Due to its enormous and wide-ranging commercial potential, the businesses and nations that control the HDTV market will eventually control the semiconductor and personal-computer markets as well.

3. *Competitor nations are pouring money into HDTV.* The Japanese government alone has targeted $700 million toward HDTV research and development. As a result, Japanese companies are expected to start selling finished units to consumers in 1991. Several Western European countries are spending more than $200 million annually to move HDTV on to assembly lines. And the advent of the European Economic Community's unification in 1992 only

adds to its potential competitive advantage.

4. *American economic health requires a manufacturing component.* Economies that are based on everybody flipping everybody else's hamburgers don't grow, because nothing of lasting value is created. That's the dangerous fallacy put forth by proponents of a service-based economy, who advocate that the U.S. should abandon manufacturing to the emerging powers of Asia, because "they can do it better." In fact, they can't do it better—if that were true, Honda, Toyota, and Nissan wouldn't be building cars in America.

No, American workers haven't lost the ability to produce quality products efficiently. Corporate leaders and elected officials in the U.S. have lost the vision to make long-term investments. They've lost the will to compete. They've lost the drive to be number one. That must change, and HDTV provides the perfect opportunity.

In addition to the military and technological need for high-definition television, HDTV has a virtually unlimited consumer market. Within the next 10 years, it will become as ubiquitous as the VCR is today.

As a result, the raw economic impact will be extraordinary. The Economic Policy Institute (EPI), a Washington-based think tank, has testified before Congress that if the U.S. captures 50 percent of domestic demand, HDTV and spin-off microelectronics sales would add up to a $10 billion trade surplus.

That alone is reason enough to warrant government support. But in addition, millions of American jobs are at stake—mostly high-paying unionized jobs requiring skilled workers and allowing a comfortable standard of living. These workers will also be able to purchase HDTV and other products, which, in turn, will create more jobs.

In recent years, the U.S. has done a better job in retaining good jobs in high-tech industries than in the manufacturing sector, where jobs have gone off shore. American superiority in the electronics industry has created over 2.5 million jobs, one million of these in the past two decades. The electronics industry employs one out of every nine manufacturing workers in the U.S., three times greater than automobile manufacturing and nine times larger than steel production. Electronics manufacturing jobs will grow substantially with a substantial U.S. investment in HDTV—but they will plummet without one.

According to EPI, if the U.S. does not develop a strong HDTV industrial base it won't just be the HDTV receiver and VCR industry that suffers. The personal computer, automated manufacturing equipment, and semiconductor industries will also deteriorate, affecting more than 750,000 jobs within 20 years. Because the HDTV industry will have an imposing impact on other industries that use electronic components, such as the automobile and telecommunications industries, the total number of jobs sacrificed to a noncompetitive HDTV industry will rise to *two million!*

Without a U.S. investment in HDTV, several foreign firms will move some of their plants to the U.S., but the jobs they create will involve low skills and low pay. The sophisticated electronic equipment, which involves highly skilled workers, will be produced abroad and shipped to the U.S. for assembly. These assembly jobs, compared to the jobs created abroad by the HDTV industry, will be low paying and will involve little skill.

The U.S. cannot afford to lose more high-paying manufacturing jobs. These jobs historically have been the backbone of America's middle class. Without a strong industrialized economic base, the schism between the rich and poor will continue to swell.

America would be shooting itself in the foot twice. A weak HDTV industry in the U.S. will not only sacrifice its future industrial base, it will lower the country's standard of living by substituting good jobs with low-paying ones.

To make matters worse, without a strong HDTV industry, American electronics will be more expensive than those from Europe and Japan. As a result of massive government investments, European and Japanese firms will begin mass production of HDTV before U.S. firms can, thus reducing the price of foreign-made electronics.

U.S. companies will have difficulty selling their high-priced HDTV products in their home markets. Orders for electronic products from abroad will increase, exacerbating problems with the U.S. trade imbalance and threatening American jobs. "Made in the USA" will become a fond memory.

Here's what government must do to put the "can-do" back in U.S. economic policy:

1. *Make the needed investments.* Even at $30 million, the DARPA investment would have been inadequate. The U.S. needs a public-private partnership matching its competitors' commitment, in order to make HDTV happen. It must not be cowed into myopic inaction by the budget deficit. This is an investment that will repay itself many times over in the future, with increased tax revenues, jobs, enhanced competitiveness, and corporate profitability.

   To those who say government has no business involving itself with business in this manner, I say this: it already does. The Economic Policy Institute has pointed out the Defense Department's role in the disintegration of America's consumer-electronics manufacturing capability. Over the past decade, high-technology firms have shifted their research, development, and manufacturing to meet the Pentagon's needs. With virtually no international competition and with favorable procurement practices, this shift has become the most profitable short-term choice.

   Unfortunately, this trend has meant an orientation toward low-volume, high-cost products, the exact opposite of what's needed in the consumer marketplace. As a result, American companies must be weaned off the Pentagon and taught to do things in reverse. Today, the real opponents of the U.S. are economic, not military. What better use of government than to take the money spent on the massive military buildup of the '80s and invest it in a massive economic buildup in the '90s—starting with HDTV?

2. *Relax antitrust laws.* Without this needed step, U.S.

companies will be fighting with one hand tied behind their backs. SEMATECH, a U.S. government–sponsored manufacturing consortium of semiconductor firms, was instrumental in maintaining U.S. sales in microchip technology and manufacturing. A similar approach, which would allow such companies as IBM and Digital to jointly develop and produce HDTV receivers, is essential.

3. *Play hardball with competitors.* The U.S. must not allow cheap, imported HDTV components to strangle its infant HDTV industry in the crib. The Commerce Department must put other nations on notice that the U.S. will take whatever actions are needed to ensure a level playing field.

In the past, America has always responded creatively and energetically to the challenges that faced it. Now, more than any other time in their modern economic history, Americans need that same response.

*Editor's note:* This chapter is an expanded version of a newspaper editorial by Mr. Barry.

# 14

# Market Forecasts and HDTV's Impact on Other Industries

*Congressional Budget Office*

*D*iscussions of U.S. policy toward high-definition television focus on three market forecasts of HDTV sales. All predict that HDTV receivers will be available to consumers in the early 1990s and that programming for HDTV will begin shortly thereafter. The following summarizes the three market forecasts:

- The National Telecommunications and Information Administration (NTIA) commissioned a report, commonly known as the Darby report, which examines the growth patterns in sales of previous consumer electronic products and forecasts the growth of HDTV.[1] The Darby report differs from the others in that it has both low-growth and high-growth scenarios. It forecasts sales of HDTV receivers and videocassette recorders (VCRs) for the years 1997 through 2008. It concludes that if HDTV replicates the growth pattern of color televisions and VCRs, the market could expand rapidly to achieve annual sales of 18.6 million units by the year 2008.
- The Electronic Industries Association (EIA), a trade association that represents all electronics companies in the United States (whether foreign- or U.S.-based), commissioned a study of the HDTV market.

That study predicts how the projected demand for HDTV will affect U.S. employment and overall economic activity.[2] Unlike the other two reports, the EIA forecast, which covers the period 1989 through 2003, excludes VCRs for HDTV. The EIA report concludes that sales of HDTV could reach 13.1 million units by 2003, raising income and employment in the electronics sector, regardless of whether the market is supplied by foreign- or domestic-based producers.

- The American Electronics Association (AEA)—a trade association that, in the area of HDTV, represents U.S.-based firms—is mainly concerned with how demand for HDTV could affect the competitiveness of U.S. high-technology industries.[3] The AEA forecast for both HDTV receivers and VCRs (and affiliated industries) covers the period 1990 through 2010. The AEA report forecasts sales of 11 million units by 2010 and concludes that unless 50 percent of the U.S. HDTV market is supplied by U.S.-based firms, the market shares of U.S. producers in other industries, most notably semiconductors and personal computers, will decline substantially.

One problem with any analysis of competitiveness is the definition of the nationality of firms and the products they make. Most analyses characterize the nationality of a firm by the nationality of its stockholders. This definition, however, ignores two important criteria: where production and where research and development (R&D) occur. It is not clear why a product made in Mexico by a firm owned by U.S. citizens is a U.S. product, while a product made in the United States by a firm owned by foreigners is a foreign product. R&D presents an equally vexing problem: While U.S. policy makers might want all firms to perform their R&D in the United States, many U.S.-based firms find it advantageous to perform their R&D in Japan or Europe. While recognizing these limitations, this report defines U.S.-based firms as firms owned by U.S. citizens, while foreign-based firms have foreign ownership.

The Congressional Budget Office has compared

forecasts in terms of annual sales, market value, and prices. Most data in these reports were presented in annual terms, but some data were given as cumulative figures. To facilitate analysis and comparison, CBO converted the cumulative data into yearly data.

The forecasts of unit sales provide a range of market size and timing. Figure 14-1 compares the unit sales forecasts of the three reports. Darby's high-growth and the EIA forecasts see the market for HDTV sets becoming as large within the next 15 to 20 years as the market for color television sets was in the early 1980s (over 10 million units per year), with EIA being much more sanguine about HDTV's market growth.

Since color television and VCRs have been among the most successful consumer items introduced over the last several decades, these forecasts of HDTV sales are probably near the high end of the potential range of market size. By contrast, Darby's low-growth scenario illustrates what would happen if HDTV ended up serving only a specialized (or niche) market, such as that which projection television—which is most commonly found in commercial establishments rather than in private homes—serves today.[3] In this scenario, HDTV receiver sales would remain at the 100,000- to 150,000-unit level, less than half the current sales of projection television.

All forecasts of market value, except Darby's low-growth scenario, suggest that the market value of HDTV sales will grow rapidly. Figure 14-2 presents the sales forecasts of the three reports in 1988 dollars. The data include the sales of HDTV receivers and HDTV-related VCRs, except for the EIA forecast, which excludes VCRs. Darby's high-growth scenario rises almost continuously; the others show markets that would peak at about the size of the current U.S. market for VCRs and television sets, which totaled about $11.5 billion in 1987.[5] The AEA forecast for HDTV and associated VCRs is $10.9 billion in the year 2010; the Darby high-growth scenario is $16.2 billion in 2008.[6] The EIA's forecast of $11.7 billion in 2003 for HDTV receivers, but not VCRs, is comparatively high next to the AEA forecast and Darby's high-growth scenario for both receivers and VCRs.[7] By contrast, the

**Figure 14-1
Projected Annual U.S. Unit Sales of High-Definition Television Receivers**

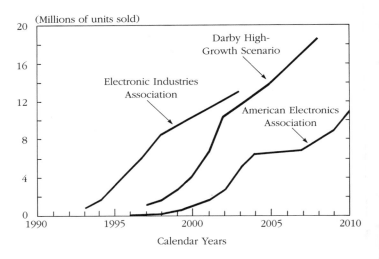

*Sources:* Congressional Budget Office, based on data from the Electronic Industries Association, "Television Manufacturing in the United States: Economic Contributions—Past, Present, and Future" (produced under contract by Robert R. Nathan Associates, Washington, DC, February 1989); Department of Commerce, National Telecommunications and Information Administration, "Economic Potential of Advanced Television Products" (prepared for NTIA by Darby Associates, April 7, 1988), referred to as the Darby Report; American Electronics Association, "High Definition Television (HDTV): Economic Analysis of Impact" (prepared by the ATV Task Force Economic Impact Team of the AEA, Santa Clara, CA, November 1988).

*Note:* The unit sales in the Darby low-growth scenario (not shown) are less than 150,000 per year.

**Figure 14-2**
**Projected Annual U.S. Market Value of High-Definition Television Receivers and Related Video-cassette Recorders**

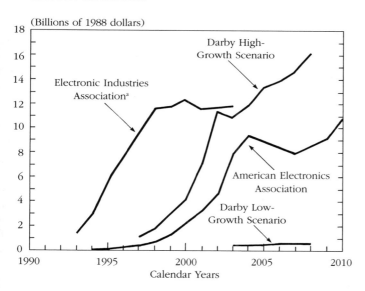

(Billions of 1988 dollars)

*Sources:* Congressional Budget Office, based on data from the Electronic Industries Association, "Television Manufacturing in the United States: Economic Contributions—Past, Present, and Future" (produced under contract by Robert R. Nathan Associates, Washington, DC, February 1989); Department of Commerce, National Telecommunications and Information Administration, "Economic Potential of Advanced Television Products" (prepared for NTIA by Darby Associates, April 7, 1988), referred to as the Darby Report; American Electronics Association, "High Definition Television (HDTV): Economic Analysis of Impact" (prepared by the ATV Task Force Economic Impact Team of the AEA, Santa Clara, CA, November 1988).

a    Excludes VCRs

Darby low-growth forecast suggests a modest market of less than $600 million in 2008.

Underlying these forecasts of market value and unit sales are very different assumptions about the pattern of price and cost reductions. Darby assumes that to achieve a high rate of market growth, the prices of HDTV receivers would have to be only a few hundred dollars more than current sets. In the Darby low-growth scenario, HDTV receiver prices remain at about $2,500. By contrast, both AEA and EIA have HDTV enjoying a multimillion-unit market with prices well above $1,000, almost triple the 1989 average price of a standard color TV set ($350) and a multiple of the projected value of improved-definition television—an intermediate technology closer to standard color television sets. The EIA and AEA forecasts of market growth would be roughly consistent with the history of color television: In the mid-1960s, the yearly demand for color television sets grew from 700,000 units to five million units while color television prices were roughly constant at $1,300 in 1988 dollars.[8] At that time, the differential between monochrome receivers and color receivers was $840 (in 1988 dollars).

While a forecast of a new product's sales is subject to a high degree of uncertainty, there is reason to believe that these analyses overstate how rapidly the market will actually grow. CBO makes no attempt, however, to provide an alternative forecast.

The forecasts share two critical assumptions—that HDTV will be a success with consumers, and that the market will grow rapidly from the outset. None of the forecasts has a scenario for true failure, even though such products are frequently launched. Many, if not most, new consumer products are not major successes; the videophone and the ultrasound shower, for example, were forecast to be mass-consumer products.[9] Many products fail outright, while others take decades to become a major commercial success. Some fail as a result of bad design or marketing, while others fail because of demographic or other factors. In all but the Darby low-growth scenario, however, HDTV receivers are forecast not only to be successes, but to be one of the most successful consumer-electronics products of the last several decades. The timing of HDTV's success is also an issue. Considering the differences between HDTV and other successful consumer products—for instance, the VCR is a complement to color television, while the HDTV receiver is a substitute—using their history of success as a pattern for HDTV's market growth may be misleading.

*Will Consumers Buy HDTV?* The forecasts assume that consumers will prefer HDTV to conventional television. The actual evidence in this regard is mixed, and should temper the forecasts' optimism.

A study by the Massachusetts Institute of Technology (MIT), for example, gauged consumer reaction to HDTV in side-by-side comparisons with conventional color television sets. The study found that almost two-thirds of viewers preferred the HDTV sets over conventional sets, but that viewer preference was highly conditional and sensitive to such factors as program content and how close viewers were to the screen.[10] In one instance, 89 percent of viewers watching football footage preferred the conventional color set over the HDTV set when seated about 10 feet (the average home viewing distance) from 18-inch sets (the average size of home television tubes).[11] In another instance, when seated 3 feet (the ideal distance for viewing HDTV) in front of a larger (28-inch) monitor, 95 percent of viewers preferred watching the opening pageantry of the 1984 Olympics on the HDTV receiver rather than on the conventional color receiver. These strong swings in viewer preference in response to screen size, viewing distance, program content, and even color tone serve to reinforce the study's conclusion that "the preference for HDTV...is highly conditional and context dependent."[12]

Since the side-by-side comparison yielded a preference for HDTV, but not a dramatic one, the MIT study questions the claims that HDTV is a technology as revolutionary as color TV was in the 1950s. These doubts undermine the basic assumption that HDTV market growth will resemble the enthusiastic consumer response to color TV. Instead, the MIT study suggests, HDTV may replace mainly large-screen televisions—the so-called premium end of the television market.

The evidence regarding the growth of the premium end of the television market is mixed. On the one hand, analysts such as those at EIA and AEA point to the growing market share of large-screen tabletop television receivers. Television receivers with tubes 19 inches and under represented virtually all sales of tabletop and portable television receivers in 1980, but this share has dropped substantially in recent years and the share of larger-screen television receivers—those with tubes 25 inches and over—has increased quite rapidly. But this growth in market share has come largely at the expense of console television sets, whose share of the market dropped in half during this period. In addition, receivers with larger tubes were not widely available until recently. In fact, the premium end of the television market, defined as the sum of console sales and the sales of tabletop and portable receivers in the largest size category of television tube, remained constant at roughly one-quarter of all television sales during the 1980s.[13] Finally, the market for very expensive television receivers in the United States is quite small. According to rough estimates provided by EIA's Marketing Services Department, the sales of television receivers with a retail value over $2,000 approach 250,000 units per year, roughly 1 percent of all television sales.

Growth in the premium television market, however, should not be equated with the guaranteed success of HDTV. There are several types of advanced television as well as several premium versions of current television technology. The forecasts offer no reason why one specific technology—HDTV—is the logical successor to the current market leader, even if consumers ultimately accept some form of advanced television. In fact, EIA makes the implausible assumption that sales of any one type of advanced television are unrelated to sales of the other types. HDTV may have more advanced technology than improved-definition television (IDTV) and enhanced-definition television (EDTV), but in the absence of clear signals from consumers, no inferences about its likelihood of success relative to the others are reliable.

The history of television purchases makes it clear that consumers have paid substantial amounts for what they perceived as progress. However, the VCR market has also shown that technical superiority does not necessarily guarantee market success. The major VCR market share, for example, was won not by the company that had the best picture—the Beta format pioneered by Sony was widely regarded as superior—but rather by other factors, such as playing time and price, which favored the VHS format.[14] The issue is whether consumers perceive HDTV as a sufficient refinement over standard color or improved television to warrant the additional expense. Judging from the price differentials forecast by AEA, the price advantages of the older technology may be substantial, especially in the early years when HDTV will cost between $1,500 and $2,300 more than standard color television sets.

*How Fast Will HDTV Develop?* Another assumption common to all forecasts is that HDTV manufacturers are going to "get it right"—that is, produce the exact product consumers want—the first time. This assumption is questionable. In the case of the video recorder, 10 years of unsuccessful consumer-oriented video recorders preceded the introduction of the model that became an "overnight success."[15] The features that have made today's VCRs so popular were not apparent or even possible in the period when firms first decided to develop a consumer video recorder. Indeed, consumers have probably been doing some learning also, discovering what they really desire. For example, the ubiquitous personal audiocassette player is an obvious case of consumers not realizing they had a need until they actually saw the product. Similarly, the microwave oven languished for 20 years after its introduction until costs finally declined and families with two working parents found that they needed a way of preparing food rapidly.[16]

One issue that is particularly relevant to the timing of consumer purchases of HDTV is the availability of programming specifically for HDTV receivers. Color television sales, for example, did not really increase until color broadcasting became widespread. The investment decisions made by producers of television programs and other "software" that may be available for HDTV will be critical to consumers' desire to purchase the new sets, and vice versa. Any delay in programming availability, or a perception by consumers that HDTV programming is not substantially different from current programming, would probably also delay consumer interest in switching to HDTV.

The implication of these product histories and associated investment decisions about programming is that the commercial acceptance of HDTV may be a long time coming. There is no reason to assume that HDTV manufacturers will "get it right" the first time or that producers of television shows will decide to make their current production equipment obsolete by quickly shifting to HDTV production. Even if advanced television ultimately replaces the current color television system, that path may be more tortuous and may take much longer than the optimistic forecasts suggest. And, the final product may look very different from that envisioned by today's engineers.

To many industry analysts, concern about the HDTV market is but one facet of a more general concern about the U.S. semiconductor industry and other U.S. electronics industries. These concerns are evidenced primarily by the broad range of effects that HDTV is claimed to have on other segments of the overall electronics sector, and by the degree to which the relatively small HDTV market (even under the most optimistic market forecast) is supposed to influence activities in much larger markets. In reviewing the claims for HDTV made in the market forecasts, CBO found, in general, that either the markets were unlikely to be big enough to have the desired effects, or the sequences of events asserted by the studies were not sufficiently developed to warrant the conclusions that were drawn.

Supporters generally argue that HDTV sales could create an enormous demand for U.S.-manufactured inputs, and that U.S.-based firms would become more competitive as a result of supplying products to HDTV manufacturers. In most cases, CBO found that the ability of potential suppliers of semiconductors, flat-panel displays, and computers to compete in international markets would probably not be determined by the success or failure of the HDTV market. Rather, changes in U.S. market share in most of the supplier markets are more likely to result from the actions of firms and the development of technologies within those markets than from pressures or advantages created by HDTV.

The market forecasts of the three studies are used to address two different, but related, economic questions: How would HDTV affect U.S. employment and output? And, how would HDTV affect the competitiveness of the U.S. electronics sector?

The AEA report seeks to show how the presence of U.S.-based firms in a successful HDTV market would affect their ability to compete in the markets of other electronic goods. The EIA report attempts to determine how the U.S. economy—value added and employment, in particular—would be affected by a change in the combination of products that would result from a large HDTV market. The Darby report provides only illustrative statistics of HDTV's effect on domestic output and employment and does not make specific predictions about the effects of HDTV on the competitiveness of the U.S. electronics sector. Moreover, Darby does not examine the consequences of the low-growth scenario, in which HDTV becomes a niche product.

In the AEA report, the HDTV market is valued not only for its contribution to overall employment and income but because, AEA argues, it is big enough to help U.S. electronics firms advance their technology and become competitive in other areas, and it involves the development of particular technologies that will serve as "technology drivers" for the rest of the sector. Implicit in the AEA analysis is some combination of economies of scale and learning curves that results in cost reductions as sales rise (even in related markets), and a network of common technologies and components that makes the knowledge gained in HDTV markets applicable throughout the electronics sector.

The AEA report assumes that the U.S. market shares in semiconductors, personal computers, and other electronics industries are directly related to the U.S. market share in the HDTV industry: The lower the U.S. share in the HDTV market, the lower will be its market share in these other industries.[17] It also assumes that if the U.S. participation in HDTV is just slightly smaller than the U.S. participation in conventional television is now—roughly 13 percent of the world market—U.S. semiconductor manufacturing firms could lose up to 50 percent of their world market share. Even if the U.S. market position in HDTV is more than double the current U.S. market position in conventional television, AEA assumes that U.S. semiconductor firms would lose 20 percent of their world market share. Only if U.S. firms capture at least 50 percent of the U.S. HDTV market can U.S. semiconductor firms be assured of retaining their current share of the world semiconductor market. The AEA forecast assumes similar, though sometimes smaller, declines in world market share of U.S.-based producers of personal computers and automated manufacturing equipment, depending on the positions of U.S.-based firms in the U.S. HDTV market.

The EIA analysis argues that regardless of whether the firms that make the HDTV receivers are foreign- or domestic-based, the bulk of the economic effect—on employment and value added—will occur in the United States. Forecasting from current trends, the report states that over 90 percent of the HDTV receivers destined for the U.S. market will be produced in the United States and that roughly 85 percent of the value of those units will be of U.S. origin.[18] This domestic production would increase U.S. value added by $6.2 billion, relative to what it would have been if HDTV technology was not commercialized. This increase in U.S. output would raise employment by 130,000 jobs through secondary effects.[19] The Darby high-growth scenario presents a similar increase of 240,000 jobs.[20]

The EIA's rationale for the large effect on U.S. value added is based on the fact that almost all television sets with screens larger than 20 inches are manufactured in the United States, albeit predominantly by foreign-based firms.[21] Since the benefits of HDTV are best realized in larger sets, the U.S. HDTV market is likely to be dominated by larger sets and hence largely served by local manufacturing plants, regardless of ownership. Because they are fragile and bulky, large glass tubes are expensive to ship. Consequently, television manufacturers are increasingly locating their glass-tube plants in the United States, even though the electronic components within the tubes are often imported. Furthermore, a more recent EIA-sponsored study found that, in 1987, the average domestic content of a color television receiver produced in the United States was 70 percent, and the share was over 80 percent for larger sets.[22]

The three reports discussed here use their forecasts of HDTV sales to generate additional predictions. They discuss the effect of HDTV on the output and employment of the U.S. economy. They also consider the effects of HDTV on the technology, competitiveness, and market shares of U.S.-based firms that manufacture electronic components and equipment.

The forecasts of output and employment in the electronics sector are derived from input-output analysis and industry rules of thumb, which have both advantages and limitations. Economists differ regarding the usefulness of these methods to forecast changes in output and employment. They agree that in an economy at full employment—where most analysts feel the U.S. economy to be in 1989—gains in one sector can come only at the expense of losses in others. But they disagree both on the potential for overall economic gains if the economy is at less than full employment, and on the possible contributions to the economy that might occur from an accelerated shift of resources away from more stagnant sectors and toward more dynamic sectors. Rather than focus on speculative effects that are probably not unique to HDTV, this section examines the more tangible effects that HDTV manufacturing might have on the competitiveness of U.S.-based firms.

According to the rationale behind the AEA forecast, the technology developed to serve the HDTV receiver market will allow foreign firms that have this technology to use it to become dominant in other electronics markets, from personal computers to communications equipment to workstations. The advantages of the high-resolution displays and serving the massive home market for HDTV receivers overwhelms whatever technical advantages U.S.-based firms might have within these product categories. The loss of the electronic goods markets might also subsequently entail a loss of the markets for the components used in those goods, especially semiconductors.

This possible sequence of events is not only forecast by the AEA, but is widely believed within the electronics industry. In its report, however, the AEA does not show how advantages in one area (display or imaging technologies, for example) confer such overwhelming advantages to all other markets for electronic goods. In the one market it does discuss—personal computers—this foreign advantage in display technology results in a 50-percent decline in the world market share of U.S.-based producers. Thus, the technical advantage would have to be great, though AEA offers no evidence that such an advantage would occur.

A market as small as that forecast for HDTV is not likely to affect substantially the U.S. market share in computers, communications equipment, and other electronic equipment. These markets are already large and continue to grow rapidly, and they will probably remain far more significant factors than HDTV in the evolution of electronics technology. While the AEA report forecasts a $28.5 billion world market for HDTV receivers and VCRs by 2010, other U.S. electronics industries are already much larger and are likely to continue to grow at a significant pace over the next two decades.[23] Taken together, the world electronic equipment sector grew by $54 billion in 1988 to reach $461 billion.[24] Thus, not only is this sector more than 15 times larger than the HDTV market, but also its annual growth is almost double that projected for the HDTV market.

While the effects of HDTV on the commercial development of the electronics sector may be quite limited, some analysts feel that consumer electronics as a whole (of which HDTV would be a part) plays an important role in the creation and maintenance of a competitive electronics sector.[25] First, the consumer-electronics market is sizable in its own right: About three out of every 10 integrated circuits manufactured worldwide go into consumer products, although, in the United States, the share of consumer electronics is only 15 percent.[26]

Second, there is a perception that Japanese electronics companies have used the consumer market to build up their capabilities and expertise in manufacturing semiconductors and other electronic goods. Although the technology used in consumer products is not as sophisticated as that used in other electronics markets, keeping cost down is much more important and, some argue, this ability to control costs carries over to other areas. In this regard, Japanese domination of the market for liquid crystal displays and charge-coupled devices—the light-sensitive integrated circuits that make camcorders work—is seen as having come from their strong position in the VCR, wristwatch, and other consumer markets. Similarly, surface-mounting technology, a lower-cost method of mounting components on printed circuit boards, first enjoyed widespread use in consumer electronics before being introduced into computers and other areas. On the other hand, the single largest market success of Japanese semiconductor companies—the computer memory market—has thus far had little to do with the consumer market: Japanese companies had to develop that expertise directly, by manufacturing memory devices for overseas markets. In addition, cost control is not as important in other markets: Performance or some other factor is often of much greater importance than a small reduction in purchase price.

Policy discussions concerning the influence HDTV might have on the production of other electronic goods often center on semiconductors, personal computers, and flat-panel displays.

The AEA forecast assumes that if U.S.-based firms fail to control a substantial portion of the HDTV market, U.S.-based semiconductor makers will be frozen out as suppliers to this market by the foreign HDTV producers. It also assumes that foreign producers of semiconductors selling into the HDTV market will gain vast experience and be able to reduce their costs, thus driving U.S. semiconductor makers out of even more markets. This scenario seems implausible in three main areas: The size of the demand for HDTV semiconductors, the ability of foreign semiconductor producers to transfer gains in one market into gains elsewhere, and the assumed behavior of U.S. HDTV makers.

The AEA estimates that the use of semiconductors in HDTV receivers worldwide will total $1.8 billion in 2010.[27] The same report projects a worldwide semiconductor market of $303 billion in 2010 (see Figure 14-3). Even after demand for HDTV has started to grow, semiconductors used in HDTV receivers will still represent less than 1 percent of AEA's forecast of total demand for semiconductors.

The AEA then argues that the difference between U.S. firms controlling 10 percent of U.S. HDTV sales and 50 percent of U.S. HDTV sales is the difference between realizing economies of scale or learning effects in the manufacture of semiconductors and not realizing these gains. Even assuming that U.S. firms buy only U.S. integrated circuits and foreign firms buy only foreign integrated circuits—an extreme assumption—the maximum difference between these two scenarios is $272 million. Thus, the AEA report assumes that the $272 million difference reduces U.S. production costs sufficiently to increase total U.S. semiconductor sales from $62 billion (29 percent of the world market) to $124 billion (40 percent of the world market). This assumption seems implausible. Even if the worldwide HDTV market is included, the learning effects or scale economies to be realized by additional sales of $735 million, or .2 percent, of semiconductors worldwide are extremely unlikely to be that large.[28]

The AEA figures also suggest that television receivers sold in the United States, including HDTV, will use a declining share of semiconductors over the forecast period: In 1990, receivers sold in the United States will consume 1.4 percent of world semiconductor production, falling to just .4 percent in 2010. Even under AEA's forecast for worldwide HDTV sales, the HDTV market would account for less than one percent of semiconductor sales, assuming that foreign HDTV receivers have the same semiconductor content as U.S. HDTV receivers.

**Figure 14-3**
**AEA Forecast of World Sales of Semiconductors**

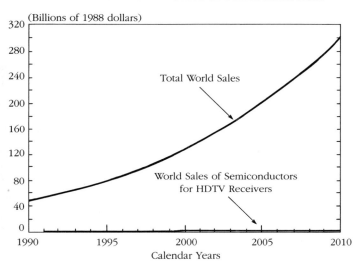

(Billions of 1988 dollars)

*Source:* Congressional Budget Office using data from American Electronics Association, "High Definition Television (HDTV): Economic Analysis of Impact" (prepared by the ATV Task Force Economic Impact Team of the AEA, Santa Clara, CA, November 1988).

*Notes:* Assumes that semiconductor content of non-U.S. HDTV receivers is the same as that of U.S. receivers. Excludes HDTV videocassette recorders.

And, in absolute terms, AEA's projections do not show that U.S. HDTV production will require significantly more semiconductors by 2010 than conventional color television will use in 1990: $682 million versus $664 million, respectively.

The AEA and others argue that, although the absolute size of the HDTV semiconductor market might be small, the integrated circuits required are so sophisticated that HDTV technology will drive engineering innovation. This seems highly unlikely for the following reasons. First, the U.S. semiconductor industry spends over $2 billion each year on R&D, and federal agencies spend another $500 million, independent of HDTV.[29] The level of HDTV R&D specifically devoted to integrated-circuit R&D is likely to be much lower than this. Assuming that all R&D is equally productive, most semiconductor engineering innovations are more likely to come from the semiconductor R&D already in place than from marginal HDTV R&D programs. Second, neither the memory devices nor the digital signal processors—the most common types of integrated circuit used in HDTV—are more sophisticated than those currently produced by the semiconductor industry. U.S. makers of integrated circuits already produce microprocessors with 1.2 million transistors each.[30] Finally, U.S. semiconductor manufacturers have traditionally led in design innovation while exhibiting weakness in manufacturing ability. This lead in design innovation is longstanding and cuts across many product categories. It seems unlikely that one program could reverse this longstanding advantage.

Although it is unlikely that HDTV receivers will drive semiconductor technology in the aggregate, they may do so for specific integrated circuit submarkets. Digital television receivers are voracious consumers of digital signal processors. Estimates suggest that this market is currently in the $500 million range, of which U.S.-based firms produce $300 million.[31] While the consumer portion of that market is only five to six percent of the total, it is growing. Thus far, however, firms with expertise in producing digital signal processors for consumer markets have not had much success in penetrating other portions of the digital signal processor market. To be made cheaply, consumer digital signal processors have to be simple and dedicated. Hence, they have very little software and little peripheral support. Whether HDTV would represent a sufficiently large market to overcome this limitation is not clear.

The other component technology frequently mentioned in connection with HDTV is the flat-panel display. The three forecasts generally assumed that cathode ray tubes would be used for most HDTV receivers; flat-screen displays are currently much smaller and much more expensive than tubes.[32] But HDTV engineers want ultimately to make the receiver displays using flat screens, which would then permit consumers to fit large receivers into their homes. Since the visual difference between HDTV and conventional color television is more noticeable with larger screens, HDTV advocates argue that flat-screen television will create both the demand and the opportunity for the HDTV market to grow. In addition, since the large-screen segment of the television market is growing, linking HDTV to a large-screen technology should improve its chances of success. The Japanese government, in conjunction with several large electronics firms, has begun a $75 million joint research effort to develop large flat-panel displays for possible use in HDTV. This highly publicized, multiyear effort has further spurred concern that U.S.-based firms will find themselves at a technological disadvantage in this market unless they initiate a similarly large research effort.

While the development of flat television receivers has been long awaited, they are just now becoming feasible. Even so, despite the Japanese research effort and DARPA funding, success is not guaranteed, as major technological and economic hurdles must be overcome. For instance, with liquid crystal displays—the kind most commonly used for flat panels—the display panel is essentially a large integrated circuit; a 12-inch screen is a 12-inch circuit. In addition, manufacturing a liquid crystal HDTV screen is the equivalent of making an integrated circuit that has 1.6 million transistors, which is as many components as the most sophisticated integrated circuits currently have. The fabrication of such massive integrated circuits at reasonable cost will be a technological challenge.[33]

If flat-screen color television becomes a reality, it will be a major step forward in consumer electronics and could affect other parts of the electronics sector. Last year, nearly 750,000 laptop computers were sold in the United States, most at premium prices. Most of these had some form of flat screen. The screens for color television need to be better in several ways than the screens for personal computers. Television receivers need full-motion video, while, outside of engineering workstations, computer monitors do not need this level of quality. Consequently, any firm that masters the technology sufficiently to produce flat color television receivers will be well positioned in the market for flat-panel computer monitors.

But, if quality flat-panel displays can be made at reasonable cost, they will come into widespread use whether HDTV becomes a reality or not. They could be used as substitutes for television tubes of any format: conventional color television or any of the three advanced television formats. In fact, small conventional television receivers that use flat-panel displays in computers are growing independently of developments in HDTV. On the other hand, if HDTV and flat-panel television succeed as a joint product, then the EIA forecast of domestic content, which is predicated on U.S. production of large picture tubes, may prove incorrect because flat panels may be easier to import than the more fragile picture tubes.

The United States, however, already imports many, if not most, of its computer monitors. According to the Department of Commerce, 8.9 million monitors, worth a total of $1.9 billion, were imported in 1988. South Korea and Taiwan together provided almost half of the total. Thus, even if flat-panel displays become much more widely accepted for use in computers, they may largely substitute for terminals the United States currently imports. In addition, the presence of so many low-cost substitutes means that Japanese producers of flat-panel displays may be unable to charge very high prices for their devices outside of specialized markets, such as the laptop computer. On the other hand, many in the U.S. industry worry about the long-term implications of relying too heavily on Japanese competitors for an input as crucial as the display, especially if it eventually supplants the cathode ray tube.

The AEA report assumes that HDTV technology will be the difference between the U.S. personal computer industry holding its current world market share (70 percent) or declining to half that level. They argue that without the incentive of HDTV technology, U.S. personal computer technology will fall behind and foreign competitors will thereby increase their market share.[34]

How realistic is this scenario? Foreign competitors

may or may not increase their share of the U.S. personal computer market, but U.S. firms already have every incentive to perform R&D and improve their manufacturing processes. The U.S. personal computer market was $23 billion in 1988, according to industry figures.[35]

The U.S. IBM-compatible market alone was roughly seven million machines in 1988.[36] By contrast, according to AEA's estimates, the U.S. HDTV market is not expected to reach that level until after 2007. Furthermore, AEA's sales projection assumes that HDTV prices will be well below $1,000 per machine, whereas the average price of the IBM-compatible personal computer is well above that. Thus, the assumption that U.S. personal-computer technology will be driven by HDTV technology requires that U.S. personal-computer manufacturing firms place more emphasis on a small market that may or may not exist in 2007, and less emphasis on a $23 billion market that exists today. Common sense and economic theory indicate the opposite.

Research and development for HDTV might be producing new designs and advances, but the U.S. computer industry is not standing still. According to the National Science Foundation, U.S. computer makers spent $8.5 billion of their own funds on R&D in 1987 alone. While not all or even most of this money was for personal computers, it indicates the level of funds available to advance related technology. Thus, unless HDTV R&D is especially productive, the major advances in computer technology will most likely come from within the industry.

Although HDTV receivers can be viewed as dedicated computers, they will not be able to run conventional software unless they are substantially modified. Most people buy personal computers because of the software. The hardware may make the software run faster or better, but the software—word processing, spreadsheets, desktop publishing—is ultimately what determines the use of the computer. Given the existing base of software—$15 million for the IBM-compatible computers alone—one would assume that most computer users would not leave their existing computers just because better displays were available on HDTV receivers. If, on the other hand, HDTV receivers incorporated personal computers that are compatible with existing software, their complexity (and cost) would be much higher, and they would be at a competitive disadvantage in both the receiver and the personal-computer markets.

When focusing on the computer market, analysts sometimes forget that the market for computer peripherals, such as printers, is also large and that the devices are complex. Many of these peripherals are almost as complex as HDTV is forecast to be, and their markets are as large as the near-term forecast for HDTV. Some of the devices now thought of as printers are, in fact, small computers. For example, the popular laser printer has a microprocessor and almost as much memory (and more can be added) as most personal computers. As semiconductor costs decline, other computer peripherals might increase in popularity. The optical scanner, for example, is waiting for cost reductions to move into the mass market. With the right software and enough memory, the scanner may be able to read magazine articles and other printed matter directly into a computer and store it in machine-readable form for later use.[37] The current optical-scanner market for computers is over $800 million per year, a figure that AEA's combined market for HDTV receivers and VCRs is not forecast to reach until about the year 2000.[38]

One technology thrust in computers, which runs parallel to but is independent from HDTV, is the development of higher-resolution displays. Computer-aided design and engineering often need high-density displays.[39] Makers of IBM-compatible computers already offer graphics systems with the same number of lines currently planned for HDTV.[40] Throughout the electronics sector, increased use of digital imaging systems as parts of larger electronic systems is widely discussed. It is quite likely, therefore, that a portion of the computer market will soon be using display systems that resemble HDTV displays. Because of this similarity, there might be some transfer of technology between the high-resolution display and HDTV. But the dynamics of that market are quite different from those of the consumer market on which these three reports focus. For one thing, these uses will largely be determined by software: Unless the design software is available, the users are not likely to demand a specific hardware system. In addition, these premium displays are usually purchased as part of an equipment system and so are only one factor in the decision to purchase the whole system. In any event, these uses will initially be of interest to a limited segment of the market.

The three forecasts of the potential HDTV market discussed in this chapter are very optimistic with regard to market size and certainly with regard to timing, although some consumer-electronics products have enjoyed the level of success they project.

Evaluation of forecasts of HDTV's effect on other electronics markets seems clearer. Even the optimistic forecasts of the HDTV market are small relative to the other electronics markets. Thus, it is unlikely that HDTV will by itself revitalize the U.S. electronics sector.

These conclusions do not necessarily mean that federal support for HDTV development is without merit. Some might support HDTV for reasons other than competitiveness in the electronics sector—for instance, employment, national prestige, or scientific advancement. The Department of Defense has a long history of pursuing display technologies that have possible military uses. In addition, HDTV development, although small on its own, might have a role to play in a broader strategy for the U.S. electronics sector.

## Endnotes

1. Department of Commerce, National Telecommunications and Information Administration, "Economic Potential of Advanced Television Products" (prepared for NTIA by Darby Associates, Washington, D.C., April 7, 1988), referred to hereafter as the Darby report. Even though the report discusses all advanced television products and makes no distinctions in the text, the author has told CBO that the scenarios are for HDTV.
2. Electronic Industries Association, "Television Manufacturing in the United States: Economic Contributions—Past, Present, and Future" (produced under contract by Robert R. Nathan

Associates, Washington, D.C., February 1989), referred to hereafter as the EIA report.

3. American Electronics Association, "High Definition Television (HDTV): Economic Analysis of Impact" (prepared by the ATV Task Force Economic Impact Team of the AEA, Santa Clara, CA, November 1988), referred to hereafter as the AEA report. The report also presents forecasts for the world HDTV market. CBO did not analyze that portion of the AEA forecast.

4. See the Darby report, p. 32. In contrast with the high-growth scenario, the low-growth scenario is undeveloped and had to be reconstructed by CBO. The Darby report states the low-growth scenario is an illustration of a possible range of sales rather than an actual forecast.

5. Electronic Industries Association, *Electronic Market Data Book, 1988* (Washington, D.C.: 1988), pp. 18 and 22.

6. The Darby report has two value trajectories for the high-growth scenario. CBO used the mean of the two.

7. CBO deflated the EIA value forecasts using EIA's projected average inflation of 3.5 percent per year.

8. Calculated from Electronics Industries Association, *Electronic Market Data Book*: 1988, various pages.

9. One analysis of new product successes and failures suggests that overly optimistic forecasts for new technology products outnumber accurate ones by between 4 and 7 to 1. See Steven P. Schnaars, *Megamistakes: Forecasting and the Myth of Rapid Technological Change* (New York: The Free Press, 1989).

10. W. Russell Neuman, "The Mass Audience Looks at HDTV: An Early Experiment" (paper presented at the National Association of Broadcasters' Annual Convention, Las Vegas, Nevada, April 11, 1988). Neuman is an Associate Professor, Audience Research Group, Advanced Television Research Project, The Media Laboratory, Massachusetts Institute of Technology.

11. According to the MIT Media Laboratory, there were some technical problems with the conventional color tape of the football game, which lowered the amount of detail available but increased the contrast. Since the level of contrast on television is lower than both nature and cinema, the unusual availability of contrast may have influenced viewers' decisions.

12. Neuman, "The Mass Audience Looks at HDTV," p. 6.

13. Data supplied by EIA Marketing Services Department. The categories of the tube sizes shifted each year. In 1980, the largest category was 20 inches and over, while in 1988 the largest category was 28 inches and over.

14. James Lardner, *Fast Forward: Hollywood, the Japanese, and*

*the VCR Wars* (New York: New American Library, 1988), pp. 151-162.

15. Richard S. Rosenboom and Michael A. Cusumano, "Technological Pioneering and Competitive Advantage: The Birth of the VCR Industry," *California Management Review* (Summer 1987), pp. 51-76.

16. Schnaars, *Megamistakes: Forecasting and the Myth of Rapid Technological Change*, p. 110.

17. The ATV Task Force of the AEA made these assumptions about declining market share of U.S. firms based on the opinions of their panel of executives and engineers from the electronics industry.

18. The 85-percent domestic content assumes that 100 percent of the electronic components, other than the tube, are imported. All tubes are assumed to be of U.S. origin. Should U.S. electronic components such as proprietary microprocessors be incorporated, the U.S. content would rise. Conversely, foreign subassembly might reduce the U.S. content. See the EIA report, "Television Manufacturing in the United States: Economic Contributions—Past, Present, and Future," pp. 50 and 68.

19. EIA report, pp. 73-76.

20. Darby report, "Economic Potential of Advanced Television Products," p. 43.

21. EIA report, p. 1.

22. Electronic Industries Association, *Consumer Electronics, HDTV and the Competitiveness of the U.S. Economy* (Washington, D.C.: February 1, 1989), p. 37.

23. AEA report, "High Definition Television (HDTV): Economic Analysis of Impact," pp. 2-4 and 2-6.

24. "1989 U.S. and Overseas Market Forecast: Looking for a Soft Landing After Hypergrowth," *Electronics* (January 1989), p. 54.

25. See Electronic Industries Association, *Consumer Electronics, HDTV and the Competitiveness of the U.S. Economy.*

26. Congressional Budget Office, *The Benefits and Risks of Federal Funding for Sematech* (September 1987), p. 19.

27. Calculated from the AEA report, pp. 2-3 and 2-10. Because AEA has no estimate for the use of semiconductors in the HDTV VCRs, this estimate understates the complete value of semiconductors used in HDTV. This and other forecasts suggest that between 6 and 12 percent of HDTV retail value is attributable to semiconductor costs. For data on the semiconductor content of HDTV, see the AEA report, pp. 2-4 and 2-10; and the EIA report, p. 50.

28. The figure of $735 million is derived by dividing $272 million (the assumed loss within the U.S. market). This

calculation assumes that all HDTV receivers worldwide have the same semiconductor content and that U.S.-based firms are going from a 10-percent to a 50-percent share of the world-wide HDTV market.

29.Congressional Budget Office, *The Benefits and Risks of Federal Funding for Sematech*, p. 43.

30.Jonah McLeod, "They're Faster, They're Denser, and They're Here: 80486 and 68040 Debut," *Electronics* (April 1989), pp. 27-28.

31.Jonah McLeod, "32-Bit Floating Point: The Birth Pangs of a New Generation," *Electronics* (April 1989), p. 72.

32.Only the AEA assumed that flat screens would account for a sizable segment of the market, and they assumed it would be at the premium end of the HDTV market. Their forecast assumes that sometime in the next five to 10 years, relatively small expensive panels of HDTV quality will appear, but that these panels will become large and less expensive over the succeeding period.

33.Rather than being a fraction of an inch square, a flat-panel display would be much larger, increasing the opportunities for flaws in the materials and errors in manufacturing. Since part of the success in making integrated circuits more dense and more cheaply has come from reducing their size, conventional processing techniques will be unusable. See Samuel Weber, "It's a New Age for LDCs," *Electronics* (May 1989), pp. 96-100.

34.AEA further assumes that the U.S. market share in personal computers could be cut in half by 1990—a date when its forecast suggests a total U.S. HDTV market of only 1,000 sets—unless the federal government and U.S. manufacturing firms aggressively pursue HDTV.

35.International Data Corporation, cited in Karen Blumenthal and Robert Tomsho, "An IBM Tagalong Sets Independent Course, with Plenty of Risks," *Wall Street Journal*, April 21, 1989, p. A1.

36.Michael Alexander, "Users Calm About Clone Battle," *Computer World*, March 27, 1989, p. 43.

37.Roberta Furger, "A Technology Comes of Age," *Infoworld*, June 12, 1989, pp. 47-51. One type of scanner—an optical character reader—needs an average of 4 megabytes of memory, exactly the size projected for an HDTV receiver.

38."The PC Juggernaut Bogs Down," *Electronics* (January 1989), p. 59.

39.Tom Manuel, "PC Graphics," *Electronics* (April 1989), pp. 84-88.

40.Bob Ponting, "Beyond VGA," *Infoworld*, June 26, 1989, pp. 45-49.

*Editor's note:* This chapter was excerpted from "The Scope of the High Definition Television Market and Its Implications for Competitiveness," a staff working paper from the Congressional Budget Office, published July, 1989.

POLITICS AND TECHNOLOGY

# 15

## The Politics of Technology:
## Congress's Role in HDTV

*Rep. Edward Markey*
*D-Massachusetts*

*E*ach day I find myself more and more distracted
by the headlines. Extra. Extra. Read all about
it. "Japan invades Hollywood." "Greenspan
goes to Moscow."

Today's headlines tell us that the most stunning
political transformation since World War II is taking
place right now in the Soviet Union and in Eastern
Europe. In this decade, the bipolar philosophy of the
Cold War, pitting communism against capitalism, is
being replaced by perestroika. And telecommunications—much of it pioneered by Ameri-cans—has
been at the heart of this change.

The change was brought home to me not long ago
as I went out to eat one night in Dubrovnik, Yugoslavia.
There were a number of restaurants located down an
alley with the maître d's standing outside hawking their
dishes. This one was selling steak and that one goulash.
When one of the maître d's grabbed my arm and asked
me where I was from, I said, "The United States." And
he said: "Where?" And I said: "Boston." And he said:
"You guys are trying to steal our star basketball player,
Dino Radja, to play for the Celtics."

And the next thing I knew I was surrounded by three
more waiters, all arguing that Dino should be playing

with Michael Jordan in Chicago or Magic Johnson in L.A. but not with the broken-down, over-the-hill Celtics. They all agreed that Dino would never play for the Celtics. Sure enough, a short time later, the headlines in the Boston papers read: "Dino Returns to Yugoslavia."

Clearly they all knew more about what was happening in the U.S. than I knew about what was happening in Yugoslavia, thanks to their satellites, television stations, and VCRs.

My recent conversations in Eastern Europe did more than underscore the impact of technology on developed societies. It reinforced my belief that the global issues of war and peace are being pushed aside by other issues.

At the height of the Cold War, there was a certain logic to the argument that the issues of war and peace, capitalism and communism, that were being played out by surrogates in Eastern Europe, Latin America, and Southeast Asia should take precedence over the national economic interest. Regardless of how you feel about that Cold War era, whether you'll miss it or whether you applaud its passing, that era is over. When Alan Greenspan travels to the Soviet Union to give out fiscal advice like an economic CARE package and when waiters in Yugoslavia start to criticize the Celtics, you know the world is changing.

The late Harvard historian Barbara Tuchman has written in *The Pursuit of Folly* that folly is a factor playing a remarkably large role in government. In her view, folly consists of assessing a situation in terms of preconceived notions while ignoring any evidence to the contrary.

Consider that, in the interests of Cold War economic policy, the U.S. refused to act during the 15 years that Japanese companies were dumping television sets in the American market. Then, when the U.S. government finally did take sides, in the interest of Cold War politics, it took the side of the Japanese. The U.S. was willing to subordinate its commercial interests to the Cold War and, as a consequence, its prosecution of the TV dumping case was not as vigorous as it could have been or should have been.

While America's traditional foes—the Soviet Union and the Eastern Bloc—are paralyzed by unprecedented internal and economic turmoil, its traditional allies—Japan and Western Europe—have already begun an assault on the world marketplace.

Already the U.S.'s competitors have recognized what it has not: that the resources that defined success in the Industrial Age—labor and natural resources—have given way to the resources of the Information Age—knowledge and ideas. It is clear that telecommunications and associated technologies could very well be to the future what coal has been to the past and oil is to the present—the essential fuel of the 21st century. Foreign competitors have recognized the shifting global economic and political forces and have seized the initiative in a number of industries—especially in telecommunications.

Both Japan and Western Europe, for example, are responding to this brave new world by investing billions of dollars in emerging technologies such as high-definition television, supercomputers, optical-fiber networks, and artificial intelligence.

The Japanese and the Europeans have a plan to direct their public and private sectors in order to promote their success in these emerging technologies. This plan even involves television programs; programmers are already anticipating the time in 1992 when the 12 nations of the European Community become one market of 323 million people. So now they have a cop program that has a leading character from each of the EC countries, redubbed into every EC language, that in some countries is now more popular than "Miami Vice."

The important policy question in Sony's purchase of Columbia Pictures, then, is not whether foreigners control the broadcast rights to beloved American cultural icons such as *The Big Chill*, "Who's the Boss," or "Wheel of Fortune." The Sony-Columbia deal does, however, even at its most benign, raise questions about the impact of foreign-controlled, vertically integrated companies on the U.S. Sony's Akio Morita has already noted (in *The Washington Post*, October 4, 1989), for example, that his purchase of Columbia was calculated to produce entertainment software to go with Sony's electronic hardware products—movies for its older products as well as programming for its new products such as its Video Walkman (a hand-held VCR/TV combination). What is more, Morita has made no secret that he believes the Columbia purchase will allow Sony the kind of synergy that will give it a critical technological leg up on the competition with regard to such future technologies as high-definition television.

Contrast this with the recent actions of Zenith, the only American-owned television manufacturing company. It is nothing short of appalling that Zenith was confronted with the choice of either HDTV or computers. According to a recent study, there are a minimum of five Japanese companies (and probably an equally large number of European companies) with vertically integrated abilities to manufacture electronic products such as semiconductors, laptops, mainframes, and both heavy and consumer electronics. Then look again at Zenith. Instead of being in a position to take advantage of the synergies between its computers and its television technologies, it has been forced to sell off its computer business to compete in HDTV. As a result, Zenith is facing the economic reenactment of *Bambi Meets Godzilla.*

A worst-case scenario would portray foreign investment increasing up to a point in the U.S. where it could potentially influence the American telecommunications network. How could this happen? Say a company like Sony used its combination of programming and new products and its distribution system of movie theaters to whet the appetites of U.S. consumers for new technologies such as HDTV, which would in turn dominate the marketplace.

Many people, including the FCC and the Congress, have been working hard to preserve an HDTV standard that will not make existing television sets obsolete. Yet all our best efforts could be voided by Sony, if it gains the market power to force its own standard on the American marketplace. That could happen *if* Sony converts its newly acquired 2,700-title film library to the Japanese HDTV standard; *if* it rents these films on discs that can only be viewed on Japanese hardware; and *if* the American consumer begins to demand these superior products. Under this scenario, Sony could tear up U.S. regulatory plans.

With this in mind, the U.S. must construct a new way of looking at the world that reconciles these new economic realities. Somehow it must construct a set of policies that, instead of keeping erstwhile allies out of American markets, give the U.S. a fair chance of taking them on. And somehow, America must construct a flexible regulatory system that will allow it to cope with the ferocious change that is occurring in the telecommunications industry.

If it can come up with a comprehensive telecommunications strategy, the U.S. will be able to avoid the first law of holes. That is, once you find yourself in one, you stop digging any deeper.

I was reminded of the speed of change in the industry recently when Northern Telecom unveiled its new family of switching and transmission products that clear the way for telephone companies to put picture phones in every home and bring studio-quality video to businesses. Perhaps the 21st century is closer than we think.

The trouble is that the U.S. is going through something of an identity crisis. Ever since the 19th-century French author Alexis de Tocqueville travelled America, it has been known as a nation that worships change. "America is a land of wonders," he wrote in *Democracy in America*, "in which everything is in constant motion and every change seems an improvement." Today, Americans are beset with what I call technological ancestor worship. They still want to seek out the future but have to protect everything in their past. Americans are rapidly losing their claim to being the New World while what they used to think of as the Old World is poised to pass them by.

Think of how Americans watch the Olympics. They do not especially care about how other nations are doing. And they do not especially care about how the individual athletes from other nations are doing. Just give them one good American hero or heroine, one Mary Lou Retton story, and they are satisfied. This global athletic competition is no different from the global economic competition. But the U.S. is approaching the technological future as if the Japanese and Europeans are in the eighth inning of a tightly fought ball game while the Americans are still arguing over the best way to get to the ball park.

Even the Pentagon acknowledges this. In its 1989 report on Soviet Military Power, for example, it notes that "it is somewhat ironic that, although the Soviet Union constitutes the greatest threat to U.S. security, the greatest challenge to the U.S. technology and industrial base will almost certainly come from the United States' own allies." The report also notes that "in 1970 the U.S., Germany and Japan each invested about one percent of their gross national product in research and development. Yet by 1989, the Japanese and the Germans had increased this rate by 50 percent or more, while the U.S. remained at one percent."

It is the height of folly for the U.S. to be handing its foremost economic competitors the keys to the industries that have the potential to dominate the 21st century. But the U.S. is lagging far behind both Japan and Western Europe in the research, development, and production of a number of key communications technologies.

Chief among these is high-definition television. Those who would dismiss this new technology as solely an entertainment medium miss the point. This new high-tech hardware, whether it is used for HDTV or some other application, will most certainly have a fundamental impact on how we go about obtaining information.

Foreign competitors accept this as an article of faith. In the summer of 1989, Takashi Fujio, the head of Matsushita's Electric Hi-Vision Center, said that "HDTV is an infrastructure which will influence the entire image processing industry; this technology will hold sway over a nation's industrial power and destiny."

In my view, the U.S. could be in danger of being held in sway by its competitors as long as it continues to steadfastly ignore that its economic future is tied directly to its ability to transform itself into something like an Electronic States of America. In my view the U.S. ignores this new electronic technology at its own peril.  In the same way that it lost automobiles in the '60s, television sets in the '70s, and VCRs in the '80s, America is threatened with losing the key products of the next century.

How many more new technologies can the U.S. afford to lose before it begins to affect its standard of living, its ability to create jobs, and the future of its children?  Without an investment on both the public and private levels, Americans face the real prospect of a diminished nation in the next century.

Just as interstate highways and railroads were the lifeline of the old industrial economy, sophisticated communications networks will be essential to economic growth in the Electronic States of America.  Competitors have recognized this new reality and are moving. Japan's Nippon Telephone and Telegraph Company has embarked on a massive $240 billion capital improvement program aimed at bringing integrated network services to every business in every city in Japan by the early 1990s.  Not to be outdone, France's government-owned telephone company has invested almost $2.5 billion since 1981 in network improvements, including the free distribution of millions of "minitel" terminals to consumers across that country.

Meanwhile, the U.S. has no coherent plan for bringing sophisticated voice, data, and video services to the American marketplace.  While the Bush administration has recently unveiled a $2 billion program for super-computer research and a high-speed data network, the program would be limited to linking elite university professors and Fortune 500 companies.  It is the equivalent of constructing an eight-lane superhighway only for those who can afford to drive Rolls Royces.

The U.S. prides itself on a telecommunications system that provides service for all—rich and poor, urban and rural, big business and small business.  As America moves into an era when information will be its most important commodity, it is even more important that it protects this principle—not only for the sake of equity, but for the sake of promoting overall economic growth and opportunity.  Providing economic opportunity for all is what has distinguished America above all other nations. It can afford to be no less democratic in the Information Age.

The U.S. needs to treat these emerging technologies in a comprehensive manner, and to offer an integrated approach that coordinates regulatory, tax, and antitrust policies. Any approach will need to include direct federal action to help leverage private-sector investments in emerging technologies and infrastructure. And it will need to include a plan to bring sophisticated voice, data, and video services, not just to U.S. elites, but to as many American homes and businesses as possible.

The U.S. must not only accept—but act on—the admonishment of Shakespeare's Caesar that "the fault is not in our stars...but in ourselves."

This admission of guilt must include the recognition that the U.S. is unable to grasp the essential link between the needs of the telecommunications industry and its own macroeconomic problems. The fact is that the U.S. has mortgaged its future, and has become the world's largest debtor nation for the worst of reasons—pursuit of short-term profits ahead of long-term growth.

Whether you call this outlook "short-term-itis" or "now now-ism," the U.S. has used its instruments of debt without regard for its responsibilities to the future. This relentless pursuit of the short-term gain diminishes the U.S.'s future as a high-tech nation. According to the Under Secretary of Defense, in a 1989 written report, such pursuit of the short-term profit "has emerged as an important reason for the lack of effectiveness of American technology-based businesses."

Furthermore, Americans must recognize the symbiotic relationship that exists between its domestic short-term focus and foreign investment. U.S. economic fates are intertwined with others as it has become increasingly interdependent on foreign investment. Foreigners now hold more than $200 billion—about one-tenth—of our nation's $2 trillion debt.

Another grievous American fault is that the U.S. has neither a vision nor a policy for its telecommunications future. The Japanese have a minister responsible for telecommunications policy. The EC governments have ministers responsible for telecommunications policy. Why doesn't the U.S. have one? It is now time for Congress—and the Bush administration—to implement a telecommunications vision.

It is up to the House Telecommunications Subcommittee to develop the comprehensive policy that can take on global competition. The emerging technologies bill is a step. Supercomputers are a step. Review of cable telcos is a possible step. And, to be sure, there will be other possible steps along the way.

Not every question about the future of an Electronic States of America has an answer, just as there may not be answers to every question provoked by perestroika or by foreign takeovers. But, if the U.S. continues to delay, waiting for the perfect answer, it will be more and more difficult to come up with possible solutions.

For the life of me I don't know how to break this thing down into an issue of liberal versus conservative or Democrat against Republican. But if we Americans don't put aside our bickering and our divisiveness, we will lose to our competitors.

Unquestionably the time has come for the U.S. to construct a cohesive telecommunications policy. Unquestionably, the world is changing and the U.S. must change with it if it is ever to evolve during the 21st century. What American doesn't need is a xenophobic cry for protection or retribution. What it does need is a telecommunications game plan that, rather than keeping erstwhile allies out of American markets, gives the U.S. a fair chance at taking them on throughout the world. I am optimistic about U.S. chances as long as we all work together with the goal of a better, stronger, and more productive America as our guide.

# 16

## Implications of HDTV on U.S. Policy

*Senator Al Gore*
*D-Tennessee*

*H*DTV has become a political dividing line, where opposing conceptions of government meet. On one side of this line are those who believe the marketplace alone should dictate the course of events in the development of new technology. On the other side are those who observe that in the real world there is no such thing as a marketplace undisturbed by government, and the issue is not whether the government should be involved, but rather the manner in which government influence is to be employed.

The stakes surrounding the debate over HDTV are plainly very high, measured in any or all of a number of dimensions. At the first level of concern is whether the United States will be a factor in the global market for HDTV and HDTV-related technologies and products. At the next level of concern is the future of the U.S. semiconductor industry and the implications for that industry of an HDTV market that could be dominated by foreign enterprises. At yet another level of concern is the future of the U.S. computer industry, a portion of which may become intertwined with HDTV as new products are developed based on a fusion of display and computer technologies.

The issue, at what might be considered the ultimate

level, is America's ability to hold its own as a force in a world economy that is increasingly dominated by the ability to process information in digital form. A global civilization is emerging, and the common language of that civilization is binary code. Any nation that occupies a strategic position across the sources, channels, and processors for that kind of information will be in a position to derive enormous economic, cultural, and political benefit.

The United States has been overwhelmingly dominant in this area for so long that it takes that dominance for granted, as if it were part of the natural resource base of the country. It did not think through a national strategy to attain its dominance. It occurred, much in the same way as the British Empire was said to have developed in the 19th century: in a fit of absent-mindedness. Its global position was the result of American leadership and advantages in a multitude of areas undergirded by its economic power and the relative technological and economic weakness of others, for the first 40 years after the end of World War II.

Today, however, that leadership is under siege from every direction. And, it is under siege by people reasoning strategically about their future in relation to that of the U.S. Their progress to date is not accidental. Any accidents that have contributed to their rapid rise toward dominance relate to errors of American government policy, and to corporate and social behavior. The plans of others and errors by the U.S. harmonize in such a way as to powerfully accelerate America's decline and others' rise.

This is not the place to recite the history of American economic folly over the last 10 years. But it is important to note that the destruction of the U.S. domestic electronics industry occurred during this period and is one of its major chapters. The next chapter may record the demise of the U.S. semiconductor and computer industries.

It is symptomatic of America's problem that the only element of the U.S. government to take serious note of HDTV in this larger context was the Defense Advanced Research Projects Agency (DARPA) in the Department of Defense—at least until the dismissal of Dr. Craig Fields

as director of DARPA in the spring of 1990. Dr. Fields'
mission included responsibility for assuring that U.S.
defense forces could draw upon a healthy domestic
industrial base for advanced technologies. Looking
ahead, he realized that military requirements for HDTV
technologies might shortly be impossible to meet from
domestic sources, given the trend toward their progressive extinction or purchase by foreign enterprises—especially the Japanese.

Dr. Fields' analysis of the situation extended to the
U.S. semiconductor industry as a whole. Its disastrous
loss of market share—in the United States and
worldwide—suggested that before long the Department
of Defense might have to rely on foreign sources for
state-of-the-art technology and products. In this connection, HDTV represented not only an example of the kind
of void that might soon exist, but also a major opportunity to prevent that void from being created.

The U.S. semiconductor industry needs a mass
domestic market if it is to have a competitive chance.
Such a market is not sufficient in and of itself. Other
factors, including the price of capital, enter the picture.
But a solid domestic market is necessary. The only way
to create that market is by leaping ahead to some
product not already dominated by foreign industries.
HDTV and HDTV-related products might well be the
only such market yet to come.

Obviously, the Department of Defense is not able
and should not attempt to create a mass market for a new
family of consumer goods and business equipment
based on HDTV. Even if it were desirable, the leverage
does not exist within the role of the department. The
Department of Defense can only try to nourish its
slender and imperiled base of private U.S. firms that are
needed to serve identifiable defense requirements. If a
broader federal effort for the creation of mass-market
potential is important to the nation, then the impetus for
that must come from somewhere else.

Initially, there was hope that help would come from
the Department of Commerce. Early in the Bush Administration, Robert Mosbacher, the Secretary of Commerce,
was a frequent and enthusiastic champion of HDTV.

There came a moment, however, when I believe he was told in blunt terms by senior members of the White House staff to drop the ball. He promptly did so, and eliminated any chance of a proactive U.S. government policy to encourage the development of a domestic HDTV industry.

Dr. Fields, however, continued to try to keep alive some specialized enterprises working in advanced HDTV display technology of potential military value. But even a residual effort of this sort created frictions within the Department of Defense and higher up in the Administration. Dr. Fields became identified as a man out of step with the philosophy of the Administration and its managers. His abrupt transfer to a position of lesser responsibility effectively decapitated support for HDTV in the Department of Defense and in the government as a whole. This was, of course, the intended result.

Consequently, the United States is alone among the major industrial states, alone among its trade rivals, in lacking a strategy for assuring its future presence in what is universally understood to be a strategic industry.

The U.S. is not out of the game yet, but the trends are not good. It is in serious danger of falling permanently behind in semiconductor manufacturing technologies. It is ironic that one of the sources of hope in this area is SEMATECH, a consortium of private companies maintained by funds in the defense budget. The irony of course is that the same Department of Defense that fails to see the logic behind HDTV as a potential new mass market for U.S.-made semiconductors is funding a consortium whose purpose is to sustain U.S. technology for semiconductor manufacturing. Any port in a storm. But where SEMATECH spends in the hundreds of millions of dollars, private corporations in other countries, often with strong governmental support, are spending billions.

The U.S. is also managing, just barely, to keep central elements of its semiconductor manufacturing capability intact. The maneuvering to salvage key elements of Perkin Elmer, which succeeded almost at the eleventh hour in preventing its sale to a foreign owner, is a case in point. Perkin Elmer found it necessary to sell its divisions that make optical lithography equipment, and since they have been one of only two American companies in this industry, a foreign takeover of Perkin Elmer would have made American chipmakers extremely dependent on foreign equipment. Fortunately, American buyers were found for this important equipment maker. The loss of Perkin Elmer's skills in the field of optical lithography would have created a gap in the domestic base capable of supporting advanced manufacturing of semiconductors. Once that happened, the odds of preventing further unraveling would decline significantly. Many other corporations that have developed specialized technologies have already been picked off, and the process continues.

Whether the United States will continue to have a complete spectrum of semiconductor manufacturing capabilities for much longer is more a matter of chance than of design. All of this represents an amazing blindness on the U.S.'s part as a nation. It is nonsense to argue that the United States should not have a definite policy in this area. It has such a policy by default. Its policy is to force its industry to pay three times the going rate in Japan for investment capital. Its policy is to negotiate for years to secure marginal changes of industrial policies in other countries, which have had the effect of savaging U.S. domestic industry in one sector after another. Its policy is to permit and even encourage business acquisitions and mergers that have no business logic, and that divert capital into debt instead of productive investment. Its policy is to permit the progressive withering of American education at every level, in the sciences, mathematics, and engineering. Its policy is to encourage the displacement of genuine investors from the stock market, in favor of institutions whose only goal is to maximize their short-term gains. It is a policy of negatives, every one of which is serving to slow the country down.

The present Administration's inflexible belief in the iron discipline of the marketplace places the theories of classical economics ahead of the plain evidence of the U.S.'s experience. America's experience is telling it that in the age of technology, nations create their own advantages: that planning, discipline, and knowledge are resources that can matter more than coal, iron, and oil. The role of government in these circumstances is to create underlying conditions that are favorable to national success in the face of vast and increasing pressure from international competition. Government does not have to overwhelm or replace market forces in order to meet this responsibility. It needs to operate with finesse to help private enterprise at critical points where private enterprise cannot help itself.

One thing government can do to help the U.S. compete for a share of the future global industry based on information processing is to get behind HDTV. Another way for government to help in this specific area is to keep funding SEMATECH and to encourage other consortia designed to help U.S. industry share the costs of remaining at the cutting edge. I believe there also is a third and very important step the U.S. government should take: Namely, it should urgently push ahead with its advantages in high-performance computer technology as a national priority.

High-performance computing is the most powerful tool available to those who, in an increasing number of fields, are operating at the frontiers of imagination and intellect. The nation that most completely assimilates high-performance computing into its economy will very likely emerge as the dominant intellectual, economic, and technological force in the next century.

High-performance computers will enable the U.S. to build more efficient engines and appliances, forecast the weather more accurately and further in advance, test new kinds of molecules with properties not found in nature for medicine and industrial materials. But high-performance computers will never be able to do all these things in the future unless we increase access through high-speed networks right away and develop the information infrastructure to realize the potential of these electronic technologies.

Legislation I have introduced is designed to provide the nation with this kind of capacity: the National High Performance Computer Technology Act. The purpose of this act is to link the nation's supercomputers into one system through a high-capacity fiber-optic network. A national network with associated supercomputers and databases will link academic researchers and industry in a national collaboration. This information infrastructure will permit the use of the nation's vast databanks as the building blocks for increasing industrial productivity, creating new products, and improving access to education. Libraries, rural schools, minority institutions, and vocational education programs will have access to the same national resources as more affluent and better known institutions.

Much of the output of this national system will ultimately be comprised of visual displays using high-definition technology. In the end, millions of users will have access to information products and services of a sort now reserved for relatively small numbers of people. The ability to place massive quantities of information at the service of the population at large, the power to manipulate that information, and the effect of HDTV as a means to fuse together forms of data transmission that are now isolated, all will combine to provide a powerful thrust into the future for America.

Others in the world are clearly aware of this potential and are moving to occupy commanding positions. Unless the U.S. acts in time to nurture its own resources, the information age will be led by others, instead of the United States. American technological supremacy, which the nation had considered as a kind of national attribute, will pass. And, so will its rewards.

# 17

## HDTV-DC:
## Washington's Concerns

*Rep. Don Ritter*
*D-Pennsylvania*

*T*he debate in Washington about HDTV is best understood as part of the larger issue of competitiveness. Competitiveness has become an important word on Capitol Hill, and with good reason. The world economy is evolving and new centers of economic power are emerging. With apparently reduced military tensions derived from the economic implosion of communism, economic power is rapidly replacing military power as the real power.

TV coverage associated with Emperor Hirohito's funeral brought home the story of Japan's incredible economic success. Many people were shocked by the blasé way both American newscasters and the Japanese referred to Japan, not only as a superpower, but as "Number One," "the number-one industrial power," "having a higher per capita income than the U.S.," and so forth.

And America's competitors are not just in the Far East. European-owned companies like Thomson and Philips now play an important role in the U.S. economy, especially in consumer electronics, and, if all goes according to plan, in 1992 the European Economic Community will become a single integrated market—larger than the U.S. market. If European products already compete in markets that America once dominated, one can only imagine what may happen after 1992.

In waking up to the Japanese and European challenge, Americans have realized that much of our electronics industry is either gone or under foreign ownership. Semiconductors, digital computers, consumer electronics, commercial TV broadcasting, and near-universal home TV set ownership were American inventions.

From a dominant position in the world market and clear technological leadership, the U.S. has gone to near insignificance in consumer electronics and in major parts of the semiconductor industry (such as manufacturing the ubiquitous computer chips called DRAMs) and even the U.S. computer industry—laptops from the low end and supercomputers from the high—is threatened.

Economists may argue that this does not matter to the U.S., that if other countries want to sell semiconductors or computers to the U.S. at prices that American manufacturers cannot match, then this is a good deal for America. But in my opinion, that argument overlooks how important it is for the U.S. to have its own mass-market electronics industry. As former Defense Advanced Research Projects Agency (DARPA) Director Craig Fields has pointed out, Americans have never wanted to let foreign interests dominate their food supply or their energy supply. These civilian industries are so vital to America's national security that the government has intervened when necessary to protect them.

In today's world, electronics technology is as important to national security as are food and energy. And of all manufacturing industries, electronics now provides the most value added per production worker. According to economics consultant Robert Cohen, the electronics sector is 25 percent more productive per worker than the auto, textile, and steel industries. This higher productivity for the electronics industry translates to the highest-wage jobs and the best standard of living for workers, neither of which America can afford to lose.

Many of us who are concerned about American

competitiveness and about the advancing sophistication of consumer electronics see high-definition TV and other high-definition systems as a potential re-entry vehicle. If the U.S. is going to get back into consumer electronics, it makes sense to do it in a new field, a new technology, a mass-market product, instead of just chasing the Japanese in products that America once had but lost.

The problem is that many experts feel the Japanese are too far ahead. They also disagree about the appeal of HDTV to the general public and the rate at which the consumer HDTV market will grow. Many in the business of advanced electronics feel, however, that HDTV will clearly be a key—perhaps *the* key—technology of the Information Age. Only enhanced visual displays, such as high-definition technology can provide, can cope with the rates at which we can now generate, accumulate, process, and transport information.

At one level, there are too many opposing groups involved in the Washington debate: On one hand there are those who feel that, if HDTV is worth getting into, American industry will see the opportunity to make money and jump in without any government action or assistance. On the other hand, there are those who feel that this technology is so important that American industry must be involved, and that government must encourage their involvement at the very least by removing the disincentives that have built up after years of bad policies.

Furthermore, people who think that the government should play a role have different ideas about what that role should be. It could range from changing the tax and antitrust laws, to direct federal funding for R&D, to subsidies or price (or trade) supports for American HDTV manufacturers.

There are many dimensions of this debate: national security, Japanese-American relations, protectionism, and the role of the government.

The government is already involved in HDTV for reasons of national security and, once again, economic security. The overall health of our technological infrastructure and defense industrial base enters into any new

199

definition of national security.

There is a long tradition of Defense Department (DoD), NASA, and National Science Foundation (NSF) funding for new electronic technologies. Opponents of a government role in HDTV point to the computer industry as one that never needed government help, but they are overlooking DARPA's critical role in the early days of computers and computer networks. The first large computers (ILIAC, for example) and computer networks (ARPANET) were paid for by DARPA before there was significant commercial interest. Other parts of the government gave the same support to communications satellites and high-performance aircraft—two products where the U.S. still leads the world. Both industries began with government money, and government interests and support still protect that lead in these vital technologies.

HDTV also has defense implications. A recent DoD report (requested by New Mexico Senator Jeff Bingaman) stressed the connections between HDTV and high-performance computing, electro-optical sensing, and advanced radar. These areas are obviously of military importance, and the U.S. can't depend on Japan and Europe to supply crucial components for them.

Because of HDTV's importance to defense, DARPA issued a broad agency announcement in December, 1988, for "the development of product and/or manufacturing technology for high definition, low cost, dynamic, multi-media displays." This was viewed by many as a stopgap effort to keep the U.S from losing out on HDTV by default. Even though only about $30 million in funding was available, the response was overwhelming; 82 formal proposals and five white papers came in as a result of the announcement. They incorporated many innovative ideas and represented companies ranging from small entrepreneurial start-ups to AT&T and IBM. DARPA identified 49 "qualified proposals," representing 160 organizations.

DARPA began awarding contracts in June of 1989. Director Fields expected that these funds would dramatically lower the barriers for U.S. companies to compete in planned FCC tests to determine the U.S.

HDTV transmission standard. It is unclear how far this funding will stretch; DARPA feels that it could effectively spend much more than the $30 million initially budgeted, but debates continue as to what the ultimate allocation and time period of DARPA's funding will be. [*Editor's note:* Subsequently, DARPA's actions have been dramatically reduced. Budget dollars have been cut and Dr. Fields was reassigned within the government and ultimately left the government agency.]

Aside from national security considerations, another dimension in the HDTV debate involves Japanese-American relations. The full story of Japan's rise to electronics superpower status has been told by Clyde Prestowitz, Jr., in *Trading Places* (Basic Books, Inc., 1988.) Prestowitz chronicles the Japanese decision to target electronics and information technology and the strategy by which they achieved dominance. He credits much of Japan's success to taking over the world dynamic random access memory (DRAM) market. In Prestowitz's words, "The Japanese knew that if they could grow faster than the Americans in the RAM segment of the market, they could become the low-cost producer of RAMs. And if they controlled RAMs, they would have taken a long step toward dominance in other semiconductors. And if they had semiconductors, semiconductor equipment, materials and everything that semiconductors went into, such as computers, would be next." From an American perspective, the Japanese strategy has been to target a new technology, invest heavily in state-of-the art manufacturing facilities, and make quality products in vast volume so as to undersell the competition and drive it out, or almost out, of business.

Richard Elkus, CEO of Prometrix, characterizes another feature of Japanese strategy as domination of the end-use markets (markets for those stand-alone products that have significant value to the individual consumer, such as video recorders, 35mm cameras, and personal computers). In Elkus's words, "Technology follows the market. If you lose your position in the end-use market, you lose your position in the supporting technology."

Because of their initial success with VCRs, the Japanese now also dominate or have a major position in professional and consumer videotape recorders, video cameras, lenses, small precision electric motors, automatic focusing systems, 35mm cameras, consumer TV receivers and video monitors, consumer and professional audio recorders, compact disk players, videodiscs, and high-speed digital-fiber transmission equipment. Each product shares technologies with the others. According to Elkus, in Japan "every technology becomes the stepping stone to the next. Every product becomes the basis for another. And the economies of scale are enormous." Elkus chaired an NSF panel established to assess the Japanese HDTV program. The panel reported that it is Japan's Ministry of Posts and Telecommunications' (MITI's) policy that Japan should no longer depend on producing computers that operate on alphanumeric characters. It must concentrate on developing computers that work with sound and sight. HDTV forms part of this strategy since, Elkus says, to the Japanese "HDTV is a market (not a technology) spanning the production, transmission, recording, processing, and display of tremendous amounts of video and audio information." Thus, to MITI, "HDTV is simply an evolution of technologies; it is not a significant breakthrough."

Elkus summarizes the American dilemma: "We face a grand strategy: The domination of interrelated end use markets, against which no single technological development or expertise can be a significant threat. . . .The goal is achievement of a predominant position in the world of advanced information systems: A market expected to dominate the 21st century—a market expected to provide a key influence on all other markets—a market expected to be the precursor to the Information Age."

The U.S. cannot counter such an overall strategy by a few simple moves, such as adopting an HDTV broadcast standard unlike any now proposed. For, in Robert Cohen's words, "Japan is a country that lives on options. They always have other alternatives because they have an infrastructure in place. Japanese spending on this infrastructure has been enormous." The U.S. needs a strategy and an infrastructure.

At one level, the current asymmetry in U.S.-Japanese trade, where the balance runs heavily in favor of Japan, is a basic part of the problem for the U.S. Japanese companies can easily enter the American market with high-quality goods that American consumers want to buy. American companies have difficulty selling in the Japanese market. Some see in this situation a deliberate strategy of *de facto* Japanese protectionism, while others attribute it to differences in the American and Japanese cultures. Cultural differences include the fact that Japanese consumers distrust imported goods and that American salespeople tend not to speak Japanese.

At another level, the past practice of Japanese businesses to sell at a loss to capture the market, as happened in DRAMs and television sets, is also a problem for the U.S. This practice is made possible by low-cost capital and efficient manufacturing. Some observers see it merely as a sharp business practice that benefits the American consumer, while others see it as dumping and illegal under American law.

At a third level, there exists for Americans the challenge of Japanese technology itself; Japan is ahead in many fields, and it wields concentrated wealth and expertise that few American companies can equal on their own.

Japanese-American economic relations are further complicated by our long-standing friendly relationship with Japan. There is tension between the desire to maintain this relationship and the problems of American companies who feel that they are being destroyed by a "Japan, Inc." that follows detrimental trade practices about which the U.S. government will do nothing.

The Japanese employ a coordinated strategy that the United States has been unable to match. Much American law and political practice is based on providing goods to the consumer at the lowest possible cost, while Japanese consumer products are usually priced higher in Japan than overseas, a practice that forces saving and capital formation. The elements of Japanese culture that put the good of the country ahead of the good of the individual encourage this. In contrast, efforts at protecting the U.S. electronics industry by limiting imports or establishing proprietary standards for HDTV broadcasting are attacked as anticonsumer. But policies that favor short-term benefits to the American consumer may be detrimental in the long term.

Our experience with DRAM chips would seem to indicate that the Japanese tend to raise export prices significantly once they have a monopoly, causing low consumer prices to disappear. But on the other hand, a valid counterexample is the VCR market, where Japanese companies compete fiercely with each other while prices drop and quality increases.

Some commentators would argue that it was American pressure that forced MITI to create a DRAM cartel and that this is the real cause of the high prices for DRAM chips. But a headline in the September 25, 1989 issue of *Electrical Engineering Times* proclaimed: "DRAM prices in free-fall." The causes for this price decrease are not clear, but they could be related to a sudden drop in demand from the computer industry, to increased production by American firms, or to the formation of U.S. Memories. (U.S. Memories was a joint venture between IBM and DEC to produce DRAM chips. It ultimately folded in January, 1990, because of its failure to get funding and support.) If the last two causes are responsible, this would support the argument that the best response we can make to the Japanese challenge is not protectionism but better American manufacturing.

Certainly simple protectionism is not the answer. This would only prop up potentially inefficient, obsolescent, and stagnant industries in the U.S. while the world passes them by—as happened with nationalized industries in Europe in the 1960s and 1970s. The American consumers' standard of living would be the big loser. No, instead of keeping others out, the U.S. must build a world-class manufacturing industry that is at the leading edge of technology and manufacturing and that can compete with anyone.

Doing this will require a national strategy in which the trade and competitiveness implications of government actions are evaluated. Major government initiatives need a competitiveness impact analysis paralleling the environmental impact statement. This was an intended result of the 1987 trade bill but so far nothing has happened.

Further action on trade is probably necessary. Particularly in electronics and telecommunications, the U.S. needs a way of enforcing a reciprocity that opens its markets to foreign competitors in some reasonable proportion to the extent their markets are open to the United States. America also needs ways to penalize egregious dumping and improper use of its intellectual property at the start—not after American companies are already out of business.

In a thought-provoking argument, Dr. Clyde Prestowitz and others point to agriculture as an example of what American manufacturing should be like. U.S. agriculture is often derided as oversubsidized and burdensome to consumers. Yet during a period when the number of farmers has dropped dramatically, U.S. agriculture is still the most efficient in the world and it competes everywhere. This situation did not happen without vigorous government activity to coordinate (some might say protect) American growers and by public-private efforts to develop new technology and transfer it to the producers.

Prestowitz points out that the U.S. Department of Agriculture has played a role like MITI's for years. This very necessary program is justified politically by invoking the image of the family farm and by emphasizing keeping America's food supply under American control. Some are saying it is time to make a similar case for America's manufacturing industry. How to do it will be an important part of Washington's political dialogue in the 1990s, and it will involve more than just HDTV.

For HDTV itself, though, what should the role of government be? Thus far government money for high-definition technology has come only from DoD and NASA, two agencies that traditionally fund new technologies. But some argue that the government should do more. The debaters do not divide along traditional liberal and conservative, Democrat and Republican lines. At one level they are divided by economic philosophy and at another they are divided by the interests of the affected industries.

The government has at least the following options for encouraging an American HDTV industry:

*Remove the disincentives that keep companies out of the HDTV business.* A major one is a U.S. tax policy that favors debt over equity, overemphasizing mergers, acquisitions, and short-term thinking at the expense of long-term investment.

*Add incentives for companies to be in the HDTV business.* These incentives might include purchases by government agencies to provide an initial market ("market seeding"), enlarging the R&D tax credit to make it a more significant factor in corporate planning, using direct funding for HDTV R&D, and implementing loan guarantees to lower the cost of capital. All of these actions would lower the risk/reward ratio in what must be viewed as a high-risk area.

*Use standard setting or other regulatory authority to delay or restrict foreign entry into the U.S. HDTV market.* The American Electronics Association, for example, recommends proprietary broadcast standards that would be licensed to foreign manufacturers at fees higher than those paid by domestic companies. But this approach is strongly opposed by broadcasters, who see it delaying their ability to engage in HDTV. And some potential American manufacturers say that whatever the standards are, foreign competitors will use them and the U.S. will end up only stunting its own growth.

At a philosophical level, some commentators feel that the government should simply stay out. Several editorials and magazine articles have labeled proposals for government funding of an industry-led partnership for HDTV research, development, and pilot production as high-tech pork-barrel programs, nothing more than pleas for subsidies. This line of argument has been particularly popular with conservative journals and organizations (*National Review* and the Heritage Foundation, for example.) The basic assertions of these organizations are: that the government cannot pick winners and losers; that HDTV may be this decade's 8-track tape player or quadraphonic sound system (in other words, that government might be propping up a money-losing technology); that HDTV might succeed in earning money, and the government should not subsidize money-making programs for private industry; and finally, that even if successful, HDTV might not have sufficient impact on the semiconductor or U.S. electronics industries to justify the government money spent.

Opponents of this view, however, argue that getting back into consumer electronics via HDTV will mean an abrupt about-face for American companies. These companies have been unwilling to invest in television, in part because government policies—America's as well as Japan's—ensure that profitability would only occur in the long term. They say that prior to the Trade Act, the U.S. had no effective mechanism to counter dumping or closed markets. Trade problems were compounded by the U.S. tax code and the comparatively high profitability of defense electronics. The government caused much of the problem and the government will have to fix it.

However, the impressive response to the DARPA initiative shows what American industry can do if given only a moderate amount of encouragement.

Opponents of a government role in HDTV frequently cite U.S. computer manufacturing as an example of a healthy industry that competes worldwide without government help and that will develop digital HDTV on its own. They feel that the Japanese are committed to an archaic analog HDTV system (MUSE) and that the U.S. is far ahead in digital electronics and fiber optics. George Gilder's widely discussed *Forbes* article "IBM TV" and his book *Microcosm* are excellent expositions of this point of view. Unfortunately, this view misunderstands both the state of our computer industry and the nature of the Japanese challenge.

Most American computer companies are systems integrators who purchase subsystems (CPUs, disk drives, memories, power supplies, etc.) and add software. They depend increasingly on overseas manufacturers for the hardware. Cray, the only surviving American supercomputer manufacturer, must buy chips from its Japanese competitors. The computer industry can thus be accurately described as "hollow." Figures from the Commerce Department show our market share for PCs declining from 75 percent to 64 percent between 1984 and 1987 and our market share for laptops dropping from 71 percent to 57 percent between 1984 and 1988. By the mid-1990s, the Japanese expect to dominate almost all parts of the computer market.

And the U.S. has no lead in digital systems or in fiber optics. According to the NSF panel quoted earlier, Japan has a three-year lead in optical fiber transmission HDTV. Working algorithms and hardware are available in Japan at 100 Mbps. The U.S. has nothing comparable. The Japanese near-term plans are for trunk distribution; long-term plans are for fiber distribution to the home.

The issue is not whether U.S. computer and electro-optics industries can develop and market HDTV if the government will just let them alone. The issue is whether America can use HDTV to re-establish a domestic consumer-oriented electronics infrastructure that will also maintain its position in computers and electro-optics. The U.S. could lose both without a significant American presence in high-definition systems.

There is also a debate among broadcasters, electronics manufacturers, and computer and information service companies. Each will be affected in a different way by what the government does or does not do in HDTV.

Television broadcasters and some cable operators emphasize the transmission issue and want to insure that HDTV will not leave them in a position like AM broadcast stations, where they are perceived as a second-class medium. Unlike almost any other country, American has a vast broadcasting infrastructure valued at some $40 billion and a tradition of "free" over-the-air broadcasting. Broadcasters feel (quoting from Congressional testimony of Joel Chaseman, CEO of Post-Newsweek Stations, Inc., and Chairman of the Association of Maximum Service Telecasters) that "indiscriminate and thoughtless development of HDTV has the potential to cripple and eventually kill America's unique local broadcast system."

Broadcasters want the FCC to mandate a single HDTV broadcast standard. They have used their own money to launch the Advanced Television Test Center to facilitate the process and are extremely wary of any kind of government action that could endanger their ability to enter the HDTV age competitively. They are not insensitive to the strategic importance of electronics manufacturing, but they argue (in the words of James C. McKinney, Chairman of the Advanced Television Systems Committee and generally a spokesman for broadcast interests) that "If we fear for the future of our semiconductor industry, then national priorities and legislation should address chip manufacturing head on."

While I understand the spirit of McKinney's statement, I don't think that the U.S. can insure its semiconductor and advanced electronics industries without reference to HDTV. The NSF panel cited earlier stated that the production model of the Japanese MUSE decoder used with their HDTV system contains 100 chips. Of these, 50 are off the shelf and 50 are application specific integrated circuits (called ASICs.) These 50 circuits represent 26 new ASIC designs. The decoder also contains 20 megabits of memory in DRAMs that are faster than anything built before. Numbers like this convince me that, in spite of some predictions to the contrary by the Congressional Budget Office, HDTV is going to drive the semiconductor industry. To paraphrase Elkus, the only way the U.S. can save its semiconductor industry is by having positions in the key end-use markets, like HDTV or computers. He believes that investing directly in semiconductors will just provide foreign competitors with better and lower-cost semiconductors to put in the products that they sell back to the U.S.

Electronics manufacturers, whose interests are different from those of broadcasters, generally support efforts to insure an American presence in HDTV research, development, and manufacturing. There are differences between electronics companies about how far the government effort should go, however, and about what, if any, restrictions should be placed on funding for U.S. subsidiaries of foreign-owned firms and about just what would constitute an American firm.

The information services and computer industries

have a third viewpoint, which places HDTV in the context of information transmission and display. They argue for a government role that would emphasize all high-resolution imaging systems and that would treat HDTV as one of many services that a national optical-fiber network would deliver. These industries generally favor building such a network first before spending money on HDTV. It is a bit of a chicken-and-egg question; opponents of this viewpoint out that the U.S. developed an automobile industry before it built an interstate highway system. It is possible that if the U.S. does not have an American presence in HDTV, then any optical network that it builds will simply interconnect Japanese HDTV terminal equipment.

Several HDTV bills have been introduced in the current session of Congress. Back in March, 1989, Congressman Mel Levine and I introduced comprehensive legislation that addresses most of the government's options. There followed a number of specialized bills that deal with single issues—antitrust, taxes, R&D funding, and so forth. Some are tied specifically to HDTV while others are aimed at strengthening American industry as a whole.

The HDTV R&D funding bills typically would authorize programs in the $100-million-per-year range through the Department of Commerce or DARPA. These bills differ in how the funds will be apportioned: Some support private-sector efforts via matching grants and loan guarantees to companies or consortia; some offer a less market-oriented effort via universities and government laboratories. (This $100 million per year is the same funding range provided for SEMATECH, a U.S. manufacturers' consortium on semiconductors.) The fate of this legislation and of the administration's policy on HDTV depends on the outcome of current debate in Washington.

Congress is receiving different signals about HDTV. Many members favor an "industry-led" effort, but industry is leading in at least three directions. Congress must weigh three good things—universal access to over-the-air broadcasting, the health of the U.S. computer industry (America's most strategic industry) and access to the information technologies of the future—and come up with a policy that reasonably protects the interests of all three while ensuring a significant American presence in HDTV development, manufacturing, and program distribution.

# 18

## Japan-Bashing

*Brian McKernan*
*Editor,*
Videography

*T*he date was September 2, 1945; the place was the quarterdeck of the battleship *Missouri*, anchored in Tokyo Bay. There, representatives of the United States, its allies, and Japan gathered to sign the instrument of surrender that would end World War II.

General Douglas MacArthur, the allied commander, closed the solemn ceremony by contrasting the devastating consequences of Japan's failed military expansionism with the allies' hopes for that Asian nation's future: "The energy of the Japanese race," MacArthur said, "if properly directed, will enable expansion vertically rather than horizontally. If the talents of the race are turned into constructive channels, the country can lift itself from its present deplorable state into a position of dignity."[1]

After the war, the U.S. spent billions of dollars rebuilding Japan and Europe. Financial aid given to Japan under the EROA (Economic Rehabilitation in Occupied Areas) and GARIOA (Government Appropriation for Relief in Occupied Areas) funds—similar to the Marshall Plan in Europe—was eventually repaid in full. In Japan, the U.S. reconstruction plan included introducing Western-style democracy, numerous social reforms, and technical and financial assistance to reestablish the country's competitive position in world trade.

These measures, the U.S. felt, were Japan's insurance against a recurrence of militarism or a possible drift toward communism.

"Compete with us now in the arena of commerce, not armed combat," was America's message to Japan. "Play the game according to our rules."

"Americans hadn't begun to realize its vast potentialities," MacArthur said of Japan in 1946 as he directed occupation efforts aimed at getting the defeated nation back on its feet. The general called the country the "springboard of the future," and 45 years later that seems to be the case, as Japan enters the '90s as an industrial superpower.

*Business Week* magazine's Global 1000 Scoreboard, an annual ranking of the world's largest companies by market capitalization, in July, 1988, listed Japanese companies as accounting for 47 percent of the value of world's top 1,000 companies, or about $3 trillion of total market value. Japan's emergence as an industrial leader, however, increasingly makes it and the U.S. at odds once again in a different kind of conflict, this one involving trade.

Trade spats are common in international commerce, and such friction between Japan and the U.S. has flared up several times in the past 30 years. Areas of contention have included: cotton in the '50s; textiles in the '60s; iron, steel, and color TVs in the '70s; and cars in the '80s. Today, however, as Japan adjusts to increased competition in these very areas from other Asian nations, expertise in high technology—semiconductors, HDTV, advanced communications, and other industries of the future—has emerged as a key to the country's economic growth in the '90s and beyond.

High-tech trade friction with the U.S. takes on added complexity because of traditional American leadership in this area, which is also associated with national security. This, combined with the continued economic success that has enabled Japan to sink more than $70 billion in the U.S. (approximately three million Americans now work for Japanese companies) and purchase choice real estate, has led to an unprecedented level of concern from U.S. government and industry leaders. When concern over the power and practices of "Japan Inc." (signifying the close cooperation between Japanese government and industry) translates into criticism, the term *Japan-bashing* has recently been applied. Just what constitutes Japan-bashing, however, seems to be a matter of perspective.

"Foreign investors own ten percent of our manufacturing base, twenty percent of our banking industry, and a third of the commercial real estate in our nation's capital," said Michael Dukakis during the 1988 presidential election. References to the economic threat from "foreigners" were a frequent theme throughout the Massachusetts Governor's bid for the White House, and his message hit home with many voters affected by foreign competition in the workplace.

Republican vice-presidential candidate Dan Quayle countered on another occasion that foreign investments in the U.S. mean jobs for Americans, but his argument wasn't helped by such well-publicized cases as the Japanese billionaire who drove around Hawaii looking at houses and sending a representative to knock on doors and make generous offers on the spot. Heavy foreign real-estate investment in that state, which threatens to make ownership of Hawaiian land too expensive for native Hawaiians, is one of many Japan-related issues with which American voters have petitioned their representatives.

Congress hears increasingly about Japan's growing economic influence and how it clashes with that of its constituents. HDTV has been a recent and dramatic example of the hue and cry that "the Japanese are taking over another industry—something must be done," and numerous politicians enjoyed much media attention in recent years as they took up the cause.

Edward Markey, chairman of the House Telecommunications Subcommittee, warned that the battle for the $50- to $250-billion market for HDTV stands to be lost if U.S. industry doesn't gear up to compete. But whether it's HDTV or some other sector of the national economic well-being, Congress frequently finds that

Japan figures into the picture. The question is, is it Japan-bashing when these elected officials conduct hearings or propose legislation to protect U.S. economic interests?

*Newsweek* magazine (October 9, 1989) featured a list of the major figures, in Congress and out, who stand on either side of the Japan-bashing fence. According to the report, on one side are traditionalists who don't want to jeopardize the U.S. strategic alliance with Tokyo and who use the term *basher* to describe Japan's critics. On the other side are those who believe the U.S. should take a tougher line on trade and economic issues. They refer to their opposition as apologists or as the "Chrysanthemum Club."

Perhaps the hardest thing is differentiating between genuine Japan-bashing and honest criticism of that country. A case in point is the *Japan Bashing Alert*, a newsletter that was published from September 1988 to April 1989 by Epistat, an international business development firm, and Portland, Oregon–based Whitman Advertising and Public Relations. The newsletter was published under contract to the Japanese advertising agency Dentsu (the world's largest) for the purpose of keeping its executives informed on any and all negative Japan-related press reports in the U.S and in Europe.

As one might expect from its title, *JBA* classified every instance of criticism of Japan as being tantamount to Japan-bashing. Comprising mostly newspaper article reprints, *JBA* closely monitored U.S.-Japanese trade friction in Washington.

"Their (Japan's) investments mean that either we end up owing them, or they end up purchasing large portions of us," the *JBA* quoted Marcy Kaptur, Democrat of Ohio, as saying on the occasion of House passage of a bill requiring disclosure of foreign investment in the U.S. (December 1, 1988). "The American people have a right to know who holds the mortgage to America."

The newsletter's February 1, 1989, edition included this reprint: " 'U.S. Secretaries of Defense and State Call for Japan to Shoulder its Share of Defense Burden.' " Elsewhere in the issue: " 'U.S. Increase in Japanese Market Share Short Lived,' . . . '43 U.S. House of Representatives members accused Japan of reneging on the trade agreement and urged the administrations to act against . . . unfair Japanese trade practices'."

The *JBA*'s wide range of articles also included everything from so dubious a bashing story as the lack of interest American audiences have for Japanese movies to the account of some genuine verbal bashing by real-estate tycoon Donald Trump. Trump, no stranger to controversy, was quoted as having stated on a TV talk show that the Japanese are "bloodsuckers who suck the lifeblood out of America."

The *JBA* notwithstanding, the ultimate "bashing" case is the tragic incident involving Vincent Chin, a Chinese-American who was beaten to death in a Detroit tavern in 1982 after an argument with a UAW member and a Chrysler Corporation foreman who reportedly mistook him for Japanese. Witnesses testified that the two men blamed him for high unemployment among U.S. auto workers.

Trade friction will no doubt continue between the U.S. and Japan as competition continues. For now the issue of Japan's lead in HDTV is one high-profile area in which current trade friction can be isolated and perhaps better understood. Numerous accusations and assumptions about Japan's lead in electronics are currently popular. It is instructive, however, to examine these claims more closely.

Claim: *The Japanese Practically Own American Industry.* In fact, the largest foreign investor in the U.S. is Britain. The second largest foreign owner of industry on our shores is The Netherlands. Japan comes in third.

Claim: *Japan Has Taken Over the Consumer-Electronics Business.* A more accurate statement should be "The U.S. lost interest in the consumer-electronics business." In 1950 there were 140 American companies making televisions in the U.S. Six years of competition later there were 100 fewer of them. Today there is only Zenith, which recently sold off its computer division to raise capital for its HDTV efforts (ironic, since the high-definition television of the future will probably be more of a computer than a television).

During the late '60s, Japanese industry did the same thing in consumer electronics that it did in many other areas of manufacturing: It made a long-term investment in producing quality products. Martin Polon, a Boston-based communications industry consultant and a contributing editor to *Television Broadcast* magazine, described the history of Japanese television success in a recent issue of that publication:

"What the Japanese did in the late '60s was to automate their television production factories, reducing by a major amount the time needed to assemble a TV set. The Japanese also concentrated on building smaller TV sets with semiconductor technology. What is perhaps the most delicious irony of all of this is the fact that the Japanese used American technology to automate their factories, redesign their TV sets, build and utilize semiconductors, and place component sets on chips."

James Lardner's *Fast Forward: Hollywood, the Japanese, and the VCR Wars* (Norton, 1987) is a well-documented history of how America's lead in such areas as videotape recording (invented by Ampex, a U.S. company, in 1956) were fumbled and surpassed by better-conceived Japanese efforts.

As for other areas of consumer electronics, in 1987 the consumer electronics divisions of venerable U.S. corporations General Electric and RCA were sold to the French giant Thomson. Magnavox and Sylvania, meanwhile, are owned by Philips, a Dutch company.

Claim: *Japan Has an Unfair Advantage in HDTV.* Advantage, yes. Unfair, no. Japanese industry began research into better television more than 10 years ago. "The reason we have been working on HDTV for more than ten years is because we are not satisfied with today's TV standard," Sony chairman Akio Morita told *Newsweek's* John Schwartz (October 9, 1989). "As scientists, as engineers, we thought we could come up with a more advanced video system."[2] It was only when HDTV research yielded actual production equipment and a variety of workable transmission schemes that a majority of U.S. manufacturers sat up and took notice. Notable exceptions were the research facilities of CBS and NBC, which had been looking into advanced and improved television for several years. Both of those facilities, incidentally, were sold off in the late '80s by corporate owners seemingly more interested in short-term returns than on long-term investment.

Perhaps the most glaring example of U.S. industry's love of short-term profit over long-term strategy was the 1985 decision by RCA to dissolve its 66-year-old broadcast division. Several years earlier, an ever-changing array of RCA management brains decided diversification was a good idea, and got the electronics company involved in rental cars, fast foods, carpets, and real estate. Several of its electronics divisions, meanwhile—businesses that RCA had nurtured for many years—were closed.

When this strategy didn't work out, corporate leadership decided to manage its defense and broadcast operations as one enterprise. When this proved unsuccessful, the parent company chose to cut expenses instead of solving its problems. Despite a stunning array of brilliant technical innovations in the late '70s, RCA finally decided to pull the plug on its broadcast division because, while it wasn't totally unprofitable, it wasn't making enough money to justify its existence.

RCA's technical innovations—including the use of CCDs (charge-coupled devices) instead of traditional tubes in portable cameras, videocassette automation systems, and "smart" studio cameras—went on to enrich the Japanese companies that were smart enough to capitalize on them. At the start of the '80s RCA was the largest exhibitor at the annual convention of the National Association of Broadcasters. By 1989 the largest exhibitor at the show was Sony, and RCA was just a memory.

Claim: *The Japanese Are Benefiting from Stolen U.S. Television Technology.* Is Japan benefiting from U.S. technology? Yes, it is. Has it stolen that technology? Well, it (usually) pays for it. Bernard Wysocki, Jr. wrote a series of articles in *The Wall Street Journal* in November, 1988, entitled "Japan Assaults The U.S. Last Bastion; Its Lead in Innovation, The Final Frontier," in which he put forth the belief that Japan is using American minds to make up for its own alleged lack of creativity in advanced technology. Wysocki cites 42 Japanese companies that are donating money to, and endowing professorships at, the Massachusetts Institute of Technology. Rockefeller University, in New York, funds $1 million worth of post-graduate researchers from Japan, but receives only $200,000 from Japan's government and industries.

Wysocki also points out that many Japanese companies establish major research and development facilities in the U.S., and hire American engineers and scientists. (Sony recently moved its Palo Alto, California, research facility and expanded it into a larger Sony Advanced Technology Research Center, in San Jose. Its budget is set at from $10 to $20 million for the next three years, and HDTV is to be main focus of its work.)

Wysocki cites cases in which U.S. companies gave away or sold technology cheaply to Japan: AT&T's transistors, RCA's television know-how, Ampex's VTRs, Honeywell's computers. Japanese companies promptly turned around and sold Americans billions of dollars worth of goods based on these technologies.

The *Japan Bashing Alert*'s issue of January 1, 1989, also cites the topic of intellectual property laws and proliferating lawsuits by small U.S. companies against giant Japanese industrial firms. "American companies are being awarded hundreds of millions of dollars in settlements against Japanese companies which pirate patented material," the *JBA* states. "Americans are obviously worried about the technological freeloading Japan has produced."

Aside from debating whether or not criticism of Japan constitutes "bashing," there are certain widely accepted aspects about that nation and its people that should be taken into account. Foremost is the obvious national will inherent in what General MacArthur called a "warrior nation" to rise from the ashes of defeat and become a world leader in four decades (albeit with a superpower's help).

Hisako Matsubara, a visiting scholar at the Hoover Institution on War, Revolution, and Peace, at Stanford University, offers these insights:

"The Japanese work ethic has its longstanding and deep roots in the religious tradition of Shinto. Work has always been in Japan 'something to the likening of the gods.' The gods were always perceived as giving their blessing to those who work. This idea comprised individual work, as well as work for the good of the community. Dedication to work and search for artistic or manual perfection are as old as Japanese history.

"If literacy in reading, writing, and calculus are any standard, Japanese education in the 18th and early 19th centuries was superior to that of any nation in Europe and superior to that of any state in the U.S.A. The degree of literacy is reported to have reached 70 percent of the total population in Japan as early as 1750, including peasants and including women. The thirst for education is not a phenomenon of today. After reopening the country in the middle of [the] last century, Japan could never have caught up with the (then still quite young) Western technology and science without a broad base of millions of educated individuals who were willing to learn and quite capable of learning."

A willingness to work hard, dedication to duty, and a desire to continually improve one's output have led to the success Japan now rightfully enjoys. Are the country's trade practices fair? The answer to that seems to be a resounding no, but then getting any nation to set aside immediate self-interest is often quite difficult. It is well documented, however, that American manufacturers have an extremely difficult time gaining access to Japanese markets. MIT economics professor Rudiger Dornbusch, writing in *The New York Times* (September 24, 1989), states: "...gaining market access in Japan is an experience not unlike attaining justice in Franz Kafka's novel *The Trial*."[3]

In addition to scrutinizing the real or imagined transgressions of Japan, the U.S. must examine its own shortcomings with even more scrutiny. American competitiveness has slipped in many areas, and the U.S.—like all great powers throughout history—seems to be poisoned by its own success. History also suggests that there is no nation so capable of progress as the United States is when determined to achieve it. But present trends don't suggest such determination currently exists.

A *New York Times* article (October 9, 1989) about a Gallup Poll of 700 college seniors shows that 60 percent had no idea when the Korean War was, 24 percent thought Columbus arrived in the New World after 1500, and 40 percent didn't know when the Civil War took place.

Time will tell if the U.S. push into HDTV will be successful or not. Sixteen of the largest U.S. electronics firms (including Hewlett-Packard, ITT, Digital Equipment Corporation, Apple, IBM, Motorola, Zenith, and Tektronix) have formed a consortium to preserve and reinvigorate the U.S. role in world electronics industry, with HDTV as the initial project. The Defense Advanced Research Projects Agency (DARPA) has committed $30 million to its own HDTV initiatives. And Zenith, NBC/Sarnoff Research Labs, and other organizations are at work devising their own advanced and high-definition television technologies.

One individual in the television equipment industry who is watching both what happens in HDTV and what his Japanese counterparts are doing is Harvey Dubner, president of Dubner Computer Systems, in Paramus, New Jersey. "I'm against Japan-bashing," states Dubner, whose company manufactures electronic graphics systems for use by television stations worldwide. He offers his perspectives on dumping—the claim that the Japanese bombard the U.S. market with below-cost goods to kill the U.S.'s own industries—and on other U.S.-Japanese trade issues.

"Every couple of years a new generation of memory chips appeared," Dubner recalls. "The new generation had increased capability and four times the capacity as the old, yet typically sold for the same price. Mostly because of price considerations, Japan was capturing most of the market. The U. S. companies complained bitterly. 'You're engaging in unfair competition, you're dumping, you can't do that.' Now there's a chip shortage, you can't ship equipment, prices don't go down.

"Dumping used to be defined as taking the last five or 10 percent of what you've got and dumping it someplace. If 50 or 60 percent of their market is the United States it can't be dumping. All we're doing is creating a situation where we do not take advantage of their low price, and we say 'Hey no, fellas, make it a high price.'

"We have much better ways of protecting our industry. If someone wants to sell me something for a low price, take it. If you *must* have an industry, let the government pay for it in one form or another, by ordering a certain amount of stuff only from American companies. That'll keep them in business.

"One thing people always said is 'The Japanese are not very good at programming,' " Dubner says, responding to a question about whether the Japanese are more innovative than the Americans. His reply refers to the belief that the allegedly lock-step, conformist culture of Japan doesn't allow for the kind of individuality and quirky creativity that has made California's Silicon Valley a cradle of computer innovation. "Somebody just went to Japan," Dubner continues, "and they came back, and said 'Watch out, we visited a company there, and we saw a bunch of programmers that were dressed in jeans and T-shirts.'

"I don't think it's a matter of innovation, of who's better, it's a matter of the Japanese getting a better education, therefore there's a higher percentage who are potentially technically productive. Success is very important to them. Of course at some point in time they'll switch around and become more like us. And then Korea will dominate. If somebody starts doing something good, I say take advantage of it. They copied us, let's copy them."

Like copying, bashing can also work both ways. Right-wing Japanese politician Shintaro Ishihara recently authored a book titled *The Japan That Can Say No,* which lambasts U.S. trade and economic policies, accuses the U.S. of trying to steal Japanese technology, urges a major Japanese military buildup, advocates favoring the Soviets in high-tech trade, and calls on Japan to assume its role as a superpower. The book has disturbed many observers, who liken its tone to the militaristic attitudes that led to Japan's initiation of war and ultimate near annihilation of itself. What has surprised many observers is that Sony president Akio Morita coauthored the book.

"Some people say I am an America basher," Morita said in a *Newsweek* account of the book's publication.

"But since I came to this country, I learned from American friends to speak frankly. Unless American industry changes its attitude, that will be a problem for all of us. The United States is the center of our free economic system, which I believe in. So for our whole free world, we need strong American leadership, and a strong American industry."[4]

## Endnotes

1. From *American Caesar: Douglas McArthur 1880-1964*, by William Manchester, © Little, Brown & Co., 1978.
2. Reprinted with permission from Sony Corporation of America.
3. Copyright ©1989 by The New York Times Company. Reprinted by permission.
4. Reprinted with permission from Sony Corporation of America.

# 19

## Fading Picture: High-Definition TV, Once A Capital Idea, Wanes In Washington

*Bob Davis*
*Staff Reporter,*
The Wall Street Journal[1]

*I*n Washington, in the spring of 1989, high-definition television was the hottest new technology in town.

U.S. consumers, the Commerce Department predicted, might well snap up $100 billion of high-definition TVs and VCRs over the next two decades. Law makers decked out Capitol Hill conference rooms with HDTV monitors to show off strikingly sharp taped footage of the Olympics. At a high-level strategy session, Commerce Secretary Robert Mosbacher and Pentagon technologist Craig Fields signaled joint efforts to promote the technology in the U.S. Then 65 officials from government, business, and academia split into three groups, war-game style, to map out an HDTV battle plan against Japan and other foreign competitors.

"It seemed possible that the U.S. government and industry could work together in high technology, as they do in Japan, to close the gap between the U.S. and Japan," says James Magid, a Wall Street analyst who specializes in electronics and attended the April, 1989, strategy session.

As of June, 1990, HDTV was in retreat, a victim of political blunders by its partisans and relentless opposition in the top levels of the Bush administration. Mr.

Fields has been ousted, Mr. Mosbacher chastened, and other HDTV allies silenced. The very term high-definition TV is being shunned in Washington for fear that any proposal identified with it will be snuffed out.

High-definition TV has come to symbolize the Bush administration's technology strategy: An open checkbook for basic research projects such as the space station or superconducting supercollider, but a tight fist for commercial technologies. The latter are best left to private investors, not government managers, says Richard Schmalensee, a member of the President's Council of Economic Advisers. If U.S. companies can't compete in high-tech markets without federal help, so be it. "There's no divine right of U.S. leadership that says our companies will always be market leaders," Mr. Schmalensee says.

The only large U.S. consumer-electronics company still in the HDTV race is Zenith Electronics Corp. Some small U.S. firms are scurrying to build high-definition equipment, but they're all desperate for money. Just constructing an HDTV manufacturing facility could cost $200 million. "It infuriates me," says Peter Brody, president of closely held Magnascreen Corp. in Pittsburgh, who pioneered high-definition technology years ago at Westinghouse Electric Corp. "The Japanese walked away with the technology, and I can't get any money" to develop new monitors, he says.

High-definition may well represent the next great leap in electronics technology. Researchers are experimenting with various technologies to create sharper images and crystalline sound, so that watching TV would become like going to a movie. In Japan, the first of the new sets are expected to cost at least $5,000 and go on sale in 1991. They would then be sold in Europe and in the U.S. in the mid-1990s. Prices would surely drop as sales increase.

Later TV sets, incorporating more advanced technology, would resemble large, thin wall screens. They could have broad applications. Wall-sized monitors could display maps and satellite images at Pentagon control centers; window-sized ones could make next-generation TV sets; smaller ones could become part of computer workstations and medical-imaging devices. "High-resolution displays will be the key to the computer of tomorrow," says Cornell economist Alan McAdams.

In the U.S., two little-known federal officials recognized the potential of HDTV and pushed the government, and a reluctant electronics industry, to back the technology in a big way. The two men couldn't be more different in temperament and political skills—and those differences helped determine their careers in government and the fate of HDTV in Washington.

The first, Alfred Sikes, headed the National Telecommunications and Information Administration, a Commerce Department backwater that he transformed into the government's first HDTV advocate. Mr. Sikes, a cautious, patient 50-year-old, once managed political campaigns in Missouri, and he approached HDTV as a backroom organizer. He recruited allies in Congress, commissioned studies showing a huge potential market, courted the secretary of commerce and pushed industry to devise a plan to pursue the HDTV market. By late 1988, more than a year after he had started working the issue, HDTV was getting hot in Washington.

Then the second official, Mr. Fields, got involved, and HDTV ignited. The 43-year-old Mr. Fields, intelligent and aggressive, was the deputy director of the Defense Advanced Research Projects Agency. DARPA, a reclusive, $1 billion Pentagon agency that has pioneered many computer technologies, sees Japan as an adversary second only to the Soviet Union. Mr. Fields viewed HDTV, which would use scads of semiconductor chips, as the coming wave in electronics—and one that Japan was determined to ride to victory.

But Mr. Fields approached the problem as a technologist, not a politician, and stumbled badly. He proposed to leapfrog the first generation of HDTV sets—costly behemoths that will use conventional cathode-ray tubes and weigh hundreds of pounds—and develop monitors as wide as four feet across but not much thicker than picture frames. This meant funding a host of exotic technologies, such as plasma gases and tiny deformable mirrors—but he couldn't find the money to pull it off.

He managed to persuade DARPA to start a largely symbolic $10-million-a-year HDTV program. Then, he clumsily lobbied Congress and other government agencies for 10 times that amount—without any luck. DARPA found itself in the worst of all situations. Its HDTV foray made it seem as if the Pentagon were trying to save the moribund U.S. consumer-electronics industry. That didn't shake loose any money. Instead, the publicity brought the attention of the new Bush White House, which chose a hands-off approach to commercial technology and, except for Commerce Secretary Mosbacher, opposed HDTV as a handout to electronics companies.

"We didn't expect that, in a friendly takeover"— Bush Republicans replacing Reagan ones—"we'd be seen as out of step," says Robert Costello, a former Pentagon official.

The usually adept American Electronics Association blundered as well. In May, 1989, the trade group was preparing a request for $1.35 billion in research grants, low-interest loans and loan guarantees for HDTV. But it didn't inform Mr. Mosbacher, until then an HDTV ally, of the specifics of the plan before he was scheduled to testify on it in Congress. Mr. Mosbacher glanced at the price tag, turned to Senate Commerce Committee Chairman Ernest Hollings, a South Carolina Democrat, and lambasted the trade group for relying on "Uncle Sugar" for handouts.

"That was the watershed event that caused HDTV to decline," Mr. Sikes says.

Any question of where the White House stood was answered around the time of the "Uncle Sugar" remark, when Mr. Mosbacher and Commerce Department Counselor Wayne Berman were invited to White House Chief of Staff John Sununu's office. Gathered around were Mr. Sununu, Vice President Dan Quayle, Treasury Secretary Nicholas Brady, Budget Director Richard Darman, Economics Adviser Michael Boskin, and other top aides. They were to discuss HDTV, and they weren't happy with the Commerce Department's role.

According to several participants, Mr. Boskin, Mr. Darman and others took turns chewing out Mr. Berman for singling out HDTV for special treatment, and they cajoled Mr. Mosbacher, a longtime friend of the president, to take a broader approach to technology. The message was clear: HDTV was out, and so was anyone who pushed it too hard.

One immediate casualty was a Commerce Department study of HDTV. The study contemplated, as an option, spending billions of dollars to support the technology and setting technical standards to block foreign competitors. "It was precisely what we go nuts about when the Japanese do it," says Mr. Schmalensee, the White House economist.

Also marked as a loser was any proposal for the government to target specific industries for low-interest loans, even if they need the aid to compete with Japanese companies that can raise capital cheaply. "If the question is who's going to get into a low-margin business [such as video displays], it's people with lots of capital who bet on high volume over the long term," says Mr. Berman, describing a slew of Japanese electronics companies but few American ones.

By summer, 1989, Mr. Mosbacher had quit the HDTV debate. "It was like a vaudeville show," says Rep. Les AuCoin, an Oregon Democrat who supports HDTV. "Someone yanked him with a hook." (Mr. Mosbacher declines comment, but a Commerce spokeswoman says he "wasn't yanked at all." Mr. Berman explains the change by commenting: "We saw that Darman was on the right track.")

Mr. Sikes didn't need a hook. Ever the adroit politician, he stopped talking about HDTV last spring and instead started campaigning successfully for the chairmanship of the Federal Communications Commission. Now Mr. Sikes, who sold a group of small-town Missouri radio stations when he joined the government, has heads of TV networks and film studios calling on him for advice. "It's a kick," he says.

He still has an important role in HDTV, too—and one that hasn't raised any flak from the White House even though it uses technical standards as a tactic against foreign competitors. He said the FCC will take until 1993 to choose a method by which HDTV programs can be broadcast here—giving U.S. companies a chance to catch up with the Japanese. It wasn't done to hurt the Japanese, "but it may have the effect of slowing them down," says Richard Wiley, a former FCC chairman who now heads the agency's HDTV advisory panel.

The FCC is leaning toward a system pushed by Zenith, say officials involved in the agency's deliberations. That would give Zenith tens of millions of dollars in royalty payments and a leg up in developing U.S. HDTV receivers.

Mr. Fields, who moved up to become DARPA's director, wanted more. He pressed for a broad HDTV technology program in which the Pentagon would seed U.S. companies with research money. He won't talk about his role, but colleagues describe him as dedicated, even obsessed, with the need to strengthen the U.S. in competition with Japan and other countries.

Mr. Fields's views closely reflect DARPA's intensely nationalistic outlook. The research agency budgets just 0.1% of its grants for foreign companies, even though many of its grants are supposedly open to all bidders. It also translated into English the Japanese best-seller *The Japan That Can Say No* so policy makers could see how the authors argue that technology can give them a military edge.

In July, 1989, when Thomas Murrin, the Commerce Department's deputy secretary, invited Mr. Fields to weekly luncheons with industry executives, the DARPA director used the occasions to press his views. "We need a national religious conversion," he wrote Mr. Murrin in August, so that the government would back heavy aid to high-tech industries, particularly electronics, without worrying whether this transgresses free-market dogma.

"Anyone who proposed any ideas for solving the competitiveness problem can be silenced by accusing him of supporting 'industrial policy,'" he complained. "And anyone who suggests an approach tailored to the unique circumstances of trade with particular countries...is a 'Japan basher.' It's hard to succeed against such a quiver of verbal arrows."

Shortly after that, when a portion of the letter appeared in this newspaper [*The Wall Street Journal*], Mr. Fields fell silent, too. He turned down interviews, avoided law makers and sent his deputy, Victor Reis, to represent him—figuring Mr. Reis wouldn't get DARPA into any more trouble because he didn't set policy.

From then on, DARPA fought a holding action against HDTV opponents. It enlisted allies in Congress to protect its tiny $10 million program. In November, 1989, these DARPA allies beat back a plan by Deputy Defense Secretary Donald Atwood to kill the HDTV program. Shortly before Memorial Day, 1989, they passed a provision that freed $20 million in additional HDTV funds the administration had tied up for six months.

But these small victories couldn't save Mr. Fields. By February, 1990, he conceded in a letter to Congress, his relations with the rest of the administration had deteriorated so badly that the Commerce Department and White House science adviser wouldn't respond to his letters asking for help in devising an HDTV policy.

The end came suddenly, in late April, 1990. An emissary from Mr. Atwood's office showed up unexpectedly at DARPA and handed Mr. Fields a piece of paper describing a new job assignment. Mr. Fields and Mr. Atwood won't discuss the incident, but congressional and Pentagon staffers say Mr. Fields was told to sign the paper, resign, or be fired for "insubordination." He signed and as of June, 1990, spends his days in a dead-end job studying the Pentagon's laboratories and looking for work outside the government. [*Editor's note:* Fields subsequently left Washington to head Microelectronics and Computer Technology Corp. (MCC), an Austin, Texas–based research consortium.]

Other HDTV advocates clammed up, too. Mr. Murrin, the deputy commerce secretary and a Westinghouse veteran, attended the joint strategy session on HDTV and once backed low-interest loans and other aid for high-tech industries. In June 1990, however, he dodges the controversy.

"It's not practical for Tom Murrin from Pittsburgh, Pa., to come here and quarterback these issues," he says. "I'll get benched so quickly you won't even see I played in the game."

But HDTV backers in Congress aren't giving up.

Some Democrats are pushing a $200 million "technology superfund" that would provide low-interest loans to high-tech companies and sharply increase funding for a new Commerce Department technology program. And Rep. AuCoin says he plans to lobby with the Defense Appropriations subcommittee on which he sits to increase Pentagon HDTV funding to $100 million. "There's life after death" for HDTV, he insists.

## Endnote

1. Reprinted by permission of *The Wall Street Journal.* ©Dow Jones & Company, Inc. 1990. All Rights Reserved Worldwide.

PRODUCTION: USING HDTV
IN THE REAL WORLD

# 20

## High Definition: Out of the Lab
## and into the Streets

*Barry Rebo*
*President and CEO,*
*REBO High Definition Studios, Inc.*

*N*ew technologies may be conceived and created in a clean, safe vacuum, but only in the
hard and gritty real world can they achieve full
development and maturity. As great as the technical and
scientific brilliance required may be to create something
as radically new and different as high definition, the
laboratory is only the beginning. Once the first HD gear
was delivered in its infant state into the hands of working
production professionals, the new medium's true potential—and problems to be overcome—began to fully take
shape. All further evolution of high-definition imaging—and it is in constant motion—is now sparked by the
feedback and insight generated by those who are
actually in the trenches.

We at REBO have always attempted to be innovators, first in current NTSC television and later in the
HDTV universe. This desire expressed itself early in the
mid-1970s, when we became one of the first video
facilities to begin to use video in many of the same ways
that film had traditionally been employed. For years, the
accepted wisdom held that the studio was video's do-
minion, and that location or field work was best reserved
for film—a limitation imposed for the most part by the
unwieldiness of video equipment, but also by the indus-

235

try's prevailing mindset at the time.

When we became one of the first facilities to acquire the early 3/4-inch U-matic equipment (the first serious professional portable video production equipment), we began to question that mindset, as well as other traditional wisdom. We started to shoot on-location, single-camera, documentary-style programs, using video in what had been the realm of film in many people's minds. What's more, we began to produce video programs on experimental subjects—an area that, again, had traditionally been home to avant-garde filmmakers. We continually attempted to push the envelope of video—to use it in a more cinematic context, but all the while attempting not to merely emulate film.

When we saw the first generation of high-definition equipment, it was immediately clear to us that it was a medium that represented a quantum leap—it had the power and the beauty of an image that could utilize much of what had been considered to be solely film's vocabulary, but at the same time it brought the power and immediacy of "real life" to an image that film could not match.

We acquired the first high-definition production system in the U.S. in 1986, shortly after seeing the equipment displayed at a trade exhibition for the first time. The impact of the image, the power of a new electronic medium that merged the "you-are-there" quality of television with the color saturation, wide-screen aspect ratio, and richness of picture available previously only with film quickly opened new vistas in video production for us. But we just as quickly became aware that, for all of HD's dazzling visual properties and space-age technology, there was a world of difference between shooting a still-life of a vase of flowers in a controlled studio environment and making a music video or shooting a theatrical feature on location in Manhattan.

In many ways, high definition presents a strange paradox—it is a sort of "lo-tech/hi-tech" medium. Many of the production and post-production tools the NTSC video community takes for granted simply do not exist in the HD universe—at least, not yet. Want to reposition an image during an HD post-production job? Forget about ADO (a digital effects unit)—there isn't any for HD. Feel that there must be a better way to tether an HD camera to the videotape recorder (VTR) than via an unwieldy and expensive length of coaxial cable? So did we. As pioneers in a new medium, we began to realize that real-life HDTV production and post-production needs required the creation of problem-solving pieces of equipment that were not being turned out by the labs building the big-ticket items (cameras, VTRs, switchers, etc.).

To address these needs, we established a subsidiary of REBO High Definition Studio called REBO Research. Interestingly, what began as an effort to supplement existing high-definition equipment has now led to a discovery of, and growing involvement in, the convergence of a whole wide range of imaging technologies, including film, video, print graphics, and computers. One reason this involvement has occurred is because we realized that in order to best develop new products to suit our production and post-production needs, we had to work with and incorporate into the problem-solving process multiple standards from all these other high-resolution imaging industries. Simply "hot-rodding" existing NTSC technology just wouldn't fill the bill if the products we developed were to be adaptable to present and future HD standards.

The research group immediately identified several key needs:

- A multiple-frame storage device for mattes, effects, and digital repositioning and manipulation of images. This would offset the lack of a high-definition ADO, while at the same time making the most of HD's incredible Ultimatte compositing capabilities and compatibility with high-resolution computer imaging.
- A compact and reasonably priced standards down-converter, capable of distributing and monitoring the HD signal in an NTSC world.
- A system to solve HD's cabling shortcomings, both in terms of the umbilical link between camera and

VTR, and the connection between the VTR and a transmission system. At the time we acquired the first system, coaxial cable was heavy, costly, and could not be extended beyond 300 feet, creating major hassles for any location shooting.

Four REBO Research products gradually emerged once these initial problems were addressed: the ReStore, an HDTV framestore based upon an Apple Macintosh computer platform and software; the ReScan, a lightweight, field-portable HDTV-to-NTSC down-converter; the ReFlect, a fiber-optic HD camera-control system; and the ReLay, a single-fiber HD optical transmission system. In each of these products, specific productions in which we were involved not only made clear that they needed to be created, but helped us in the process of actually developing them and making them work.

The initial idea for the ReStore came during the time we were working on what was to be the first American high-definition theatrical release, in 1988, a film called *White Hot*, starring Danny Aiello. Because it was a feature-length film, we often found ourselves in situations where we were going back to shoot the same scene over on different days. Obviously, this presented the usual problems in terms of maintaining continuity— matching lighting and maintaining camera-eye contact—from scene to scene. The industry's standard fix for this problem has been a pretty low-tech one: Shoot a color Polaroid of the scene, and match it the next day.

We got through *White Hot* without a hardware solution to this problem, but the seeds that led to the birth of a solution had been planted. We concluded that what was needed was a framestore unit, based on the powerful yet affordable Macintosh II computer, which would allow technical staff to finish a take, get ready for the next set-up, and literally grab one of the frames from the previous take (now stored in computer memory), and let the computer assist the crew as they go about resetting the camera and relighting the scene to keep everything matching perfectly.

Once the design of the ReStore had been initially conceived to satisfy the aforementioned requirements, it suddenly became clear that such a device could solve a multitude of other problems. For example, in order to have the capability to dynamically reposition a live or stored picture in real time, all that was required was the addition of a second frame buffer memory.

Using the Macintosh platform was advantageous for reasons other than just cost-effectiveness. We realized that the Mac's existing paint and graphics software suddenly transformed the ReStore into a high-definition graphic-paintbox-type device. Granted, it was not as powerful or fast as a dedicated graphic paintbox with HD output capability, but at about one-tenth the price, we could more than easily live with the tradeoffs.

Other ReStore applications presented themselves. For example, we realized that, because of the use of the Mac, any image that can be input into a computer (for example, via a user network) can be recorded and displayed in high definition. This capability could be useful for still-image manipulation and transmission in a wide variety of applications for industries such as medical and scientific imaging, as well as print graphics.

It also turned out that the ReStore proved to be an excellent tool for configuring 3D animation to the Macintosh, and then transferring it to high definition. In a curious reversal of the "needs-of-the-software-creating-new-hardware" scenario, this 3D capability based on new hardware eventually helped lead to the creation of an entirely new software medium—the first-ever integration of 3D computer graphics and animation with live-action HD footage. We employed it in our short film, *Infinite Escher*, starring Sean Ono Lennon, which we coproduced with directors John Sanborn and Mary Perillo, and Post Perfect, a 3D computer graphics facility.

*Infinite Escher* is a surrealistic fantasy piece, in which Sean—shot largely against a blue screen—appeared to interact convincingly within 3D computer-rendered Escher "environments" (inspired by numerous Escher paintings and drawings). It also sparked another use for the ReStore, which we have since used on other projects as well: the creation of high-contrast (Hi-Con) mattes. Most people are familiar with the famous Escher drawing depicting two hands on a sheet of paper on a tabletop, apparently drawing each other. Recreating this tableau, positioning it to appear as if from Sean's point of view, and animating it for our film was a formidable task, requiring innovation on both the production and post-production sides. With this job, as with so many we have done, no how-to manual existed.

To pull it off, we first locked off an HD camera almost straight overhead, ensuring that the camera would remain rock steady throughout the shot. The prop department then prepared a piece of drawing paper with shirt sleeves and arms drawn only up to the wrists—no hands. This paper was affixed to the top of Sean's drawing table with double-sided tape. The table was then weighted with sandbags.

Next, Sean placed his right arm over the arm drawn at the top of the paper. His wrist was made to align as closely as possible to the pencil-drawn wrist at the top of the page. We rolled tape, as Sean pretended to "draw" the lower shirt sleeve.

In order to make it appear as if Sean's right hand and wrist were emanating from a three-dimensional drawing, they were treated with heavy white make-up and black eyebrow pencil. We then placed him on the other side of the table, aligned his arm to the bottom drawing in the same way as we had done at the top. We rolled tape, and he again pretended to draw the other sleeve.

Now came the tricky part—ensuring that the illusion we were creating would indeed come off realistically. In post-production, we used the ReStore to create a high-contrast black-and-white matte. This Hi-Con divided the white left side of the screen from the black right side with an S-shaped, soft-edged line that would run above the wrist joints of both hands.

The Hi-Con was used to softly join the left half (the right-side hand drawing the left sleeve) of one shot to the right half (the left-side hand drawing the right sleeve) of the other, using a wipe (a soft-edged split screen). The result was the appearance of live hands "growing" out of pencil-drawn arms.

By using the ReStore's matte capability, we were able to relatively easily create an illusion of uncommonly powerful reality and a great deal of production value.

Had we been forced to attempt to create the same effect in film, we would have had to use more traditional animation techniques, drawing a matte on a piece of celluloid, shooting it on an animation stand, and then using it in an optical effect. This process is more costly, more time-consuming, and not necessarily more realistic than the ReStore method.

This is just one example of the kind of production and post techniques that were not necessarily envisioned by HD equipment manufacturers when they first began to build cameras and VTRs. In fact, such a technique would likely not have occurred to us until we were put in the position of having to find a way to do it, and not being willing or able to wait for the big manufacturers to come to market with a piece of gear that might help.

It is all well and good to try to shoot video with a cinematic mindset. It is quite another thing to do so with thick ropes of coaxial cable restricting your movement and generally hampering not only your freedom but your creativity.

The answer, of course, is fiber optics, something we had a great deal of experience with in our pre-HD days. The major problem: There was no fiber-optic camera-control or transmission system on the market.

As with the ReStore, it was during the shooting of *White Hot* that the need for a fiber system really began to become painfully clear. There is nothing like working on a multimillion-dollar theatrical release, using a brand-new technology not yet baptized in the field, to highlight both the advantages and disadvantages of the system with equal clarity.

The need for a fiber-optic camera-control system was pointed out to us mainly by three factors: camera mobility, RF interference (electrical interference on the screen), and camera sensitivity. While all of these are ostensibly "technical" problems, in the end they all impact heavily on the creative side. For example, there was one scene in *White Hot* where we wanted to take an across-the-street shot of the action. However, we became nervous about running thousands of dollars worth of coaxial cable across the Manhattan street on which we were shooting. This fear forced us to stop, break down, and move the entire setup across the street, affecting us financially as well as creatively.

Had the ReFlect fiber system existed at that time, we would not have thought twice about running a length of fiber cable across that street. After all, with 330 feet of coaxial costing roughly $4,000 versus an equal length of fiber costing less than $300 (not to mention being 10 times lighter and only one-third as thick), the choice is clear.

Cable length also figures heavily into camera sensitivity and picture clarity. Because the HD image begins to degrade appreciably when cable is run at too long a length, high-definition cameras are designed with "cable-compensation" circuitry to maintain signal strength. The fiber-optic system we ended up building can be run up to an unprecedented 10 kilometers and suffer no signal degradation or interference (it also is immune to RF interference—something virtually unavoidable in urban locations). This capability opens up all kinds of new creative vistas. (We were even able to shoot a recent space shuttle launch for NASA in HD, something not previously possible because, in the past, the HD VTRs could not be positioned far enough away to be safe from the intense heat of the launch.)

We also discovered another benefit of fiber-optic cable. With coaxial cable, the camera has to equalize for the amount of cable length, which causes noise. The more noise there is, the less the F-stop can be increased. When fiber-optic cable is used to replace coaxial cable, however, the camera is "fooled" into thinking that it is not attached to anything, because of the cable's light weight and purity of signal. Once this happens, there is no noise, so the camera can be pushed one F-stop faster than previously possible. This means you use fewer lights and save money. It also means that actors no longer have to labor under the harsh glare of so many hot lights, translating into better performances. Thus, here is a technical solution that spells both budgetary savings and creative benefits.

The ReLay was developed out of the same basic technology as the ReFlect, but instead of being designed as a camera-control system for production use, it is set up to act as the structural backbone of a high-resolution imaging network. Using the ReLay, any combination of high-resolution images, including print graphics and computer animation as well as closed circuit transmissions of high-definition images, can be used in the construction of multimedia production. This ability is a key requirement in the production environment of the '90s and will certainly become even more essential into the next century.

It's a funny thing—even very large companies that attempt to produce complete or "turnkey" systems for audio or video production rarely think of everything. Inevitably, they either don't build a product that real-life experience in the field proves there is a need for, or else they build one that does much more (and costs much more) than is actually needed in specific situations.

Such is the case with HDTV-to-NTSC down-conversion. Obviously, there is a real need for a broadcast-quality device that can down-convert HD signals to NTSC for transmission on today's television, and such devices are available from the major manufacturers of HDTV production equipment. However, these down-converters are extremely large and heavy, as well as expensive. We worked on a number of productions that pointed out to us the need to create a down-converter that was truly field-portable, and capable of handling such functions as creating cassettes for NTSC off-line editing, assessing lighting and composition when shooting an HDTV production also intended for NTSC release, and distributing converted HD images on a set to NTSC monitors, without the need to re-equip with additional HD monitors.

During the shooting of an HD commercial spot for O'Henry candy bars, we had one shot where we needed to frame the talent in front of a huge "OH" logo. The director was concerned about how the shot would look on an NTSC television set, both in terms of the red-and-yellow colors of the trademark, and of the framing of the shot. Since no field-portable down-converter existed at the time, the director was forced to wait for down-converts—obviously not a desirable situation. Another situation that convinced us of the need for the ReScan came during our first large multicamera music concert shoot, with Ryuichi Sakamoto at New York City's Beacon Theatre. Not only were there not enough HD monitors on the set, but the audio truck we were using at the time was equipped with an NTSC monitor, and the staff didn't want to refit for an HD monitor. These types of problems simply don't come into play, now that we have the ReScan.

Solving these varied production problems by creating all-new hardware has aided immeasurably in making our current HD productions run more smoothly and efficiently. Of course, the electronic tools are just part of the equation—there is no substitute for learning by doing.

As of June, 1990, we are in the midst of producing an ongoing, long-term music performance series. Titled "Manhattan Music Magazine" and shot live in three top New York City nightclubs, this series spotlighting eclectic musical genres has benefited greatly from the ReScan down-converter. Using the ReScan, we are now able to run NTSC feeds off our three new-generation HD cameras to the large monitor in the remote audio vehicle. This ability is essential, as the "Manhattan Music Magazine" shows are being distributed in NTSC and PAL (current television in Europe), as well as HDTV in Japan. The technical crew can thus evaluate the quality of the shots on the spot for both HD and traditional television formats.

The introduction of any new technology—especially for a creative medium like film making—is always a "chicken-and-egg" scenario. In the case of high definition this scenario is an endless cycle, where hardware development drives the creation of software, while the needs of certain types of software spur the invention of new hardware. Subsequently, this new hardware again sparks the creation of previously inconceivable software. And so on. And that's good.

We are deeply involved in a dynamic process in which high technology and creativity feed one another— where the relative vacuum of the research lab is counterbalanced by the heady urgency of the shooting stage.

# 21

## HDTV and Hollywood:
## Will the Feature Film Community
## Embrace HDTV or Resist It?

*Fern Field*
*Producer,*
*Brookfield Productions*

W hile the debate about HDTV transmission standards, broadcast standards, and now even production standards rages on in the hallowed halls of Congress and in lofty corporate executive suites, the truth of the matter is that Hollywood couldn't care less!

It is also true that major changes will have to occur—in the perception of the technology, in the accessibility of the technology, and in the physical attributes of the technology—before it is welcomed, or pursued, by American producers and production executives in our entertainment industry.

I was privileged to produce a movie utilizing HDTV in 1988 for the CBS Entertainment network. It was not my decision to do that, but I celebrated the opportunity, and looked forward to becoming acquainted with this exciting new process that was being offered to us. It is, after all, the first major change to come along in television since the advent of color TV.

We found it to be an exciting new technology—a tool we hope to use again soon. Like all new things, our use of HDTV generated some resistance and some criticisms, and required adjustments on the part of cast and crew, all of which can be minimized in the future with the proper mind-set.

However, had we not had CBS behind us, it is unlikely that, as a small production company, we would have ventured into these new waters, for fear of the unknown and for the attendant financial implications. The fact that CBS was the producing entity also made it possible for us to conduct a week's worth of tests prior to our shoot, something that would not have been economically feasible for an independent production company.

But the question remains, "Is Hollywood ready for HDTV?" Even more so, "Does Hollywood *want* HDTV?" In a random sampling of Hollywood production executives at studios and small and large independent production companies, HDTV is perceived as "premature," "very interesting, but has a ways to go," "more cumbersome than film," "more valuable in the post-production and distribution medium," and one executive said he'd gone "to a couple of meetings" and frankly his impression is "that nobody knows anything about it."

Perhaps it is different in other countries around the world, but in America change generally happens because it cannot be denied. Hollywood is not going to jump on the HDTV bandwagon until it has to.

For HDTV to be embraced by the American production community, a number of steps need to be taken. First of all, it is necessary to create a climate for change that will facilitate the perception and acceptance of HDTV. For example, a long-standing debate exists about the "film look" vs. the "tape look." It is counterproductive to feed into this discussion by claiming that HDTV will replace film. HDTV is not a replacement for film; if anything, it will replace our current 525-line videotape. What HDTV is, is a new medium, another tool that the creative community can use to express itself. Until it is perceived as such, instead of as a threat to the way most of the production crews make their living, it will have a hard time finding fans.

On the other hand, HDTV must also be economically viable if it is to compete and become known to the production community. The equipment must be available to production companies at competitive rates.

Eliminating the lab and negative costs of film does not generate enough savings to compensate for the much higher cost of HDTV equipment.

There are also some major pros and cons in the area of post-production vis-à-vis using HDTV technology. Many of our television films are now being posted on tape on a greatly accelerated post-production schedule. Movies often are completed in less than four weeks. Until the creative community as a whole stands up and says, "no, no more, we won't butcher our product that way," unrealistic (at least creatively unrealistic) delivery schedules and air dates will continue to plague those of us who make a living producing shows for television. (The latest nightmare schedule I heard of was five-and-a-half weeks from the start of photography to delivery for a two-hour television film!) This new practice portends well for HDTV because dailies can be edited overnight and viewed the next day.

In addition, since our 525-line system is the least desirable in the world, a producer is faced with the need to convert tape to film for foreign distribution, or contend with a product that is inferior and often times unacceptable in other countries. The cost of conversion back to film needs to be factored into the total project cost—something that would not happen if the project had been shot originally on HDTV.

On the other hand, Hollywood likes to show off its productions to its peers. Recently, a producer who had posted a four-hour project on tape went to the trouble and expense to make a 35mm print—just so it could be screened in a large theater in Los Angeles. (Perhaps they did not know about the large-screen video projectors that are available, or perhaps those projectors would not fit into the booth of the selected theater.) More recently, it became impossible to screen our CBS film in Los Angeles because one of the two existing HDTV large-screen projectors was in Japan being serviced and the other was in Washington for our screening there and then had to be sent to another convention.

Until equipment is available so that a film maker can show off his or her work to the community, HDTV will not gain wide acceptance in Hollywood. The new

247

Directors' Guild of America building has installed an HDTV screening room, but it will accommodate only 50 people, barely enough to hold family members of one person involved in a production. Hopefully, the booth for the Guild's largest theater will be able to accommodate a large-screen HDTV projector; hopefully the projectors will be more available in the future; and hopefully, too, they will be able to accommodate an audience of even more than 150 or 200 people.

Our HDTV dailies were gorgeous—because we were viewing them on HDTV equipment provided to us by the CBS engineering department. But the down-converted NTSC cassettes of our dailies were dismal! Had they been gorgeous, there would have been an undercurrent in Hollywood of "wait until you see this" among people who handled and viewed our cassettes. Instead, that opportunity for good word-of-mouth PR for HDTV was lost because our cassetted dailies looked worse than any dailies I've ever had converted from film negative.

We also have a beautiful 1-inch (NTSC) tape from our HDTV master. Again, the cassettes are dreadful, worse than any I've had. No one looking at a cassette of our finished film would become an advocate of HDTV. Instead they would be less inclined than ever to use it. Steps must be taken to change this.

Until the quality of the technology can be ensured and protected all the way down the line, HDTV will not gain acceptance in the American creative community.

Of course, once HDTV hits the homes and people start to look for, and clamor for, programming in this medium, Hollywood will respond to the demand. Or, if the technology becomes so much more cost effective that it cannot be ignored, then once again, Hollywood will jump on the bandwagon. But until then there is much that should be, and can be, done to arouse the interest of the Hollywood creative community, and it's not happening. The Hollywood community is not being primed to accept HDTV, so they will be the last—not the first—to begin thinking about or using this exciting new medium.

As an outsider to the debate over production standards, I find it hard to understand what all the fuss is about. To the best of my knowledge, 1125/60 is a *de facto* HDTV production standard. It's out there. It's being used. It is hard enough to get any of us to try something new. As a producer, if I decide to try HDTV, I would certainly look to the technology that has been proven in the field. Every day that goes by, 1125 experiences another day of production. Clearly, if I were to use the medium at all I would choose the medium where I can call a producer or a director who has used it and ask what the benefits and the drawbacks are. What producer is going to be the first one to try out a new HDTV production medium, when there is one that has already been tried and tested by others?

Since 1125 can be converted to 1050, or 1250, or whatever broadcast standard is finally adopted, isn't it a little foolish to be debating a production standard—when we already have one? Oh yes, I understand there is money at stake, and profits to be made, but haven't we learned, both as producers and broadcasters, that having different systems (NTSC/PAL/SECAM) has a negative impact, financially and creatively, on our productions? With the advent of more and more television coproductions between countries all over the world, isn't it obvious that we need a common production standard in HDTV, just as 35mm is the common production standard in film? Let each country determine its broadcasting standards according to its needs. But let the creative community have one production standard in this new electronic medium destined to become an integral part of our lives within just a few short years.

Still, regardless of what standard is finally determined to be the production base of HDTV, there is already a growing attitude regarding HDTV production on the part of those in the Hollywood creative community who do see HDTV in our future. The response is to *shoot on film.*

Almost to a person, the production executives I talked to felt the only way to safeguard product longevity is to shoot on 35mm film for "backward compatibility." Film processing labs are running ads about this, magazines are writing articles about it, and studios are very conscious of it.

Everybody agrees that sooner or later HDTV will be with us—if not for use, at first, in production itself, then at least in the areas of distribution, post-production, and delivery systems—and when an HDTV standard is adopted, everybody wants to be ready to convert their productions from film to the new technology. So, they are talking aspect ratios for filming with a view to the globally accepted HDTV 16:9 aspect ratio. They want to seek "pan-scanning" approval, subject to supervision from the creators of the productions, to render American products competitive, and visually pleasing, when HDTV distribution becomes a reality.

Whereas one executive mentioned that producers who are currently integrating the use of film and tape for more economical special effects would have shows that are obsolete when HDTV is available, another executive said arrangements were already being made to re-do those segments in HDTV. More and more professionals are convinced that HDTV's first impact will be felt in distribution, post-production, and delivery systems, so Hollywood is mobilizing to protect its shows from obsolescence.

On the production front there is unanimity, even among proponents of this new technology, that more work needs to be done on the equipment. Steps in the right direction are constantly occurring, but the feeling persists that more needs to be done before Hollywood will flock to produce in HDTV.

And there is no question that the rental cost of the equipment will have to come down before there is universal acceptance of HDTV as a real option to shooting on film under ordinary circumstances. The manufacturers and owners of HDTV equipment will have to find a formula that makes it feasible for even the small independent producer to consider using this new technology, if it is to become accepted universally as a new tool by the creative community.

And accept it we should. As mentioned earlier, post-production schedules for television movies are becoming nonexistent. Since this insanity is not going to stop, we might as well be thinking which of our projects lends itself to shooting on HDTV. And there is no question that anybody shooting expensive concerts and other shows on multiple camera 525-line NTSC tape is ensuring the obsolescence of their shows the minute HDTV is available.

What we, in the Hollywood community, have not yet had a chance to do is to learn how to really use this new technology to its full potential. Just as most of us only know 10 percent of what our computers can actually do—we have learned just enough about the technology to get us through our tasks—most of us do not have the mind-set yet to use HDTV to its full advantage. Since if any technology is at risk due to the advent of HDTV it is our inferior NTSC tape and not 35mm film, we should explore the benefits and potential of this new tool instead of continuing useless debate, and competitiveness, between film and HDTV.

For instance, if we can't afford to shoot at the Vatican, or the Paris Opera House, or in Moscow's Red Square, could we perhaps shoot our actors in a studio? And, could we do this less expensively than doing the same thing on film? One producer/director of photography told me that some things done on HDTV could not have been done in any other medium. We should consider HDTV an extension of our film technology and learn how to modify our thinking and approaches to this technology in order to maximize its potential, and ours.

Sadly, HDTV has become a political hot potato, with no one quite sure of the outcome. The consensus seems to be that a universal HDTV transmission standard will not be possible to achieve. Too bad. But too much seems to be at stake. And I suspect, as usual, it is the public who will suffer and wind up footing the bill.

I find it strange that some companies are asking for delays and subsidies to research and develop a technology that already exists and is out in the market and in the field, already performing what it's supposed to be doing. I understand that there's a lot of money at stake. But where were these companies and what were they doing all these years, while the technology was being developed by others? If we are willing to make concessions, and grant subsidies in this area, then why not in other areas and for other technologies and other companies?

The issue is complex indeed, and there don't seem to be any easy answers.

As late as 1988 we all felt that HDTV would start showing up in VCRs and in the home within the next few years, but I believe this new political debate and political moves in Congress have probably delayed the advent of HDTV to American homes by five to ten years. And so now, more than ever, there is almost a unanimous "wait-and-see" attitude in Hollywood, with the conviction that we will begin using HDTV as a post-production medium much sooner than as a production "tool of tomorrow."

*Editor's note:* This chapter was excerpted from a paper and presentation by Ms. Field at the 16th International Television Symposium in Montreux, Switzerland, June, 1989.

# 22

# A Sensible Look at HDTV for Motion-Picture Production and Special Effects

*Harry Mathias*
*Cinematographer*

*T*here are fundamental conflicts that result when any new equipment is introduced into the motion picture and video production communities.

The need for practicality in production equipment would seem to be obvious, but is probably the consideration that merits the most discussion, precisely because it is the one that is least often mentioned in any debate about new technology. We must be sure not to fall into the trap of thinking that the motion-picture production industry has problems that need to be solved, simply because those who advocate high-definition television have an idea that they want to try out on this industry.

Film or video production work is creative work performed under moderate to intense time and money pressures. The need to function creatively in an environment of pressure forces all members of the production community to rely on safe, proven methods. The introduction of any new equipment or new working methods into this equation threatens to upset the community's balance, even if promising ultimate advances in efficiency or creativity.

Many of those in production feel that manufacturers

253

are continually designing new equipment to solve production problems that simply don't exist. They are tolerant and even intrigued by new equipment, but they want to know that the equipment can be integrated into a production situation with a minimum of delay and disruption. They want equipment that can be operated by creative people who they already know and trust. And, they are concerned about the length of time it may take for a cameraperson to acquire "chops" on the new equipment. "Chops" is a jazz term describing a musician's familiarity with his or her instrument. If musicians spend 10 years studying the piano, they can acquire their "chops" on the piano, but then musicians don't have to contend with piano designers who rearrange the keys on their instruments every six months. This continual redesign is precisely the problem that we are facing with new production technology today, however.

One of the main characteristics of the motion-picture image—and one that has contributed to making film such a popular art form—is its tonal flexibility.

To be sure, the 35mm film image has plenty of resolution to allow it to be projected to audiences of many thousands of people on screens of 100 feet or larger in size. But, it is not film's resolution that appeals to movie audiences, it is its subtlety of tone and color. For directors or cinematographers to be able to communicate visual subtleties, they must work in a medium that is capable of reducing these subtleties. This is where video has historically fallen short and where high definition (with the same dynamic range as current video) will continue to be inadequate.

Camera sensitivity, in addition to allowing the camera to be used in low-light shooting situations, also can be an artistic consideration. Subtlety of lighting and natural reproduction of light sources has always been difficult with a camera that requires high light levels. Virtually all currently available HDTV cameras are three or more stops less sensitive than current high-speed motion-picture stocks. Certainly, in the past there were film stocks that were quite insensitive and many subtle and artistic films were able to be made on these film stocks, at the cost of much skill and trouble. But at the time cinematographers had no alternative more sensitive film stocks available. In the case of high-definition video, the available alternative is film.

It might be worth noting that cinematographers historically will not choose a less practical alternative over a more practical one, but producers may do so if lower cost is suggested as one of the advantages of the new alternative. One area of major cost-saving claims for HDTV has been in motion-picture theatrical distribution, so it is important to discuss the economic realities of HDTV.

When asked about high-definition television as a medium for theatrical display, Richard Schafer, director of market development and planning for Kodak's Motion Picture and Audiovisual Products Division, said, "two approaches have been suggested for HDTV theatrical exhibition. The first is satellite distribution of programming to theaters. A less ambitious approach would involve distribution of videotapes or discs for HDTV projection. Either one of these approaches would require an enormous investment. For example, the cost for equipping a single screen for HDTV projection has been estimated between $220,000 and $250,000 and you would still have to distribute a videotape or disc to the theater. The projected cost for converting the nation's 19,000 screens for satellite distribution of HDTV movies would be even higher, approximately $250,000 to $300,000 each. When you consider that the average annual gross revenue per screen is around $220,000, the financial incentive for exhibitors to invest in HDTV projection just doesn't seem to be there. If you took the maximum savings that could be realized by eliminating the cost of film prints, it would take decades to parlay that into recovering costs for receiving movies by satellite and displaying them electronically. And that doesn't allocate a penny for higher maintenance and operating costs for HDTV projection. All of this, of course, assumes that the public will insist on image and audio quality that is at least equal to today's 35mm quality and assumes that the HDTV system will ultimately achieve that objective."[1]

What are the alternatives? There is already a chain

of theaters in Japan projecting movies on relatively small screens in the 525-line NTSC standard. The cost for equipping those theaters for video projection is estimated to be only $30,000 to $40,000 each. In the U.S., there is a network of electronic theaters on college campuses, which receives movies and special events via satellite. This network is also based upon the 525-line NTSC system, though the image is enhanced. But whether U.S. theatrical audiences will pay for the reduced image quality of HDTV as compared to film is a question that can't be answered until it is tried.

In fact, the trend in near-future theatrical display appears to be a move toward a visual and audio experience that is an improvement over current films. ShowScan, with its 60-frame-per-second images originated on 65mm color negative film with digital sound, is just one of the large theatrical production formats on the horizon. Kodak is also seeing a rapid buildup of interest in 24-frame-per-second 65mm production and 70mm release prints. That, of course, represents a whole new set of challenges to people who are trying to electronically emulate the film look. Kodak scientists have stated that at least a 2,000-line and perhaps as much as a 4,000-line HDTV system would be necessary to produce resolution of that theatrical quality.

Another problem with using HDTV for theatrical distribution is that no suitable large-scale projection system is available for electronic images. In order to be competitive with the existing film system, the projection must put out five to ten thousand lumens at a 200:1 contrast ratio, have 1,000 by 1,800 pixels of resolution, and have an efficiency of around 1 watt per lumen. It should also be reliable, simple to operate, and require little maintenance, and the cost should not exceed $50,000.

No current technology offers these required specifications. Future projection technologies require invention in addition to engineering. However, research dollars are being spent in these areas today. Some new projection technologies, like light valve, using directly addressed liquid crystal modulators and other projection technologies using deformable mirror devices may be proven in the future.

According to the American motion-picture industry newspaper *Variety*, piracy, or unauthorized duplication of motion-picture products, is an industry whose profits reached $1 billion last year. This figure doesn't even include international piracy in the form of unreported foreign market distribution. Complex encoding schemes and the use of encryption devices promise to make motion-picture piracy a more difficult undertaking, and therefore has great appeal for producers. It certainly is a tangible argument for high-definition video distribution of dramatic films. Simultaneous worldwide distribution of high-definition films via direct-broadcast satellite also promises producers a sizeable immediate income from their products. More importantly, the profit can be realized before motion-picture piracy can significantly erode into the box office earnings of the production. This short turnaround between completion of a production and the box office return on its investment promises to stimulate production in an era of higher interest rates. It also promises in some small measure to stem the tide of rising production costs. And don't think the promise of a worldwide release before the reviews can come out is lost on the production community, either.

While quick completion of a film is important to a producer, it should also be noted that today, using standard film methods, films are being finished in as little as three and a half months, start to finish. Long delays for film completion are usually artistic in nature and not technical.

Much has been said about the advantages of HDTV for real-time special effects in motion-picture production. There are a number of reasons for this, some political, some practical. Film special effects are time consuming and costly, whereas special effects in video production are almost instantaneous and considerably cheaper. Since no satisfactory system exists as yet for projecting an HDTV image on a large screen, however, motion-picture film is suggested as the best current distribution medium for HDTV.

The cost and difficulty of motion-picture special

effects led Petro Vlahos to develop the Ultimatte (a video compositing device) in 1971, in order to streamline the labor-intensive process of motion-picture special effects. The Ultimatte could combine two film images into a composite special effect rivaling in its subtlety a composite made on film. However, the process required a method for converting film to video and video back to film with little or no loss in quality. The equipment to accomplish this conversion did not exist at that time and the idea had to be abandoned.

Years later, when HDTV was developed, motion-picture special effects was one of the first applications proposed for HDTV. Ultimatte was readily converted to the bandwidth of HDTV and the input and output devices (HDTV Telecine and an HDTV Electron Beam Recorder or Laser Film Recorder) were part of the original NHK research and development project. But even with these essential pieces put into the puzzle, there are many obstacles to overcome before HDTV can be successfully used for motion-picture special effects.

For example, the process of combining multiple film elements into a composite optical has always caused degradation of the film image. And the degradation betrays to the audience the fact that a special effect was used. If, for example, you are producing a sequence of a spaceship landing on a planet in the film *Star Wars*, you don't want the image to turn grainy or contrasty when you cut to the spaceship landing, or it will reveal to the audience that the spaceship isn't real. For this reason 70mm film is used for the special-effects elements in a 35mm motion picture. The excessive amount of quality inherent in the 70mm film compensates for the degradation inherent in the optical process.

For HDTV to be usable for motion-picture special effects, it would have to be superior to, not just equal to, 35mm film in image quality. However, even assuming that video has an advantage over film in its reduced image degradation through multiple generations of duplication, and assuming no loss in quality due to video equipment misadjustment (a very generous assumption), video still cannot achieve the image quality available in film today.

Dynamic range or film tonal range is also very important in terms of special effects. If the video-originated special effect has less dynamic range than the film into which it is cut, then it calls attention to the special effects, making them appear unreal.

In addition to its special effects usage, 70mm film is worth discussing in another context. Seventy-millimeter film is a major technological breakthrough in film system performance. A quantum leap in image quality above 35mm film, 70mm film has been used by the film industry before to combat market threats from video. Many current features, while produced on 35mm film, are released on 70mm film.

In 1959, when the motion-picture industry was feeling competition from the broadcast television industry in the United States, many pictures were filmed in 65mm and released in 70mm. Some were even shot at 30 frames per second, in a 65mm system called Todd-Ao. *Ben Hur* was filmed in an anamorphic 65mm system known as MGM Camera 65, and projected in Panavision 70, a 70mm format. Eastman Kodak reports increased interest recently in 65mm camera negative stock, but no increased sales to date. Panavision Inc. reports a great deal of interest in 65mm cameras and, for the first time in 20 years, has developed a new 65mm camera. Several productions are, as of June, 1990, utilizing 65mm film. It is presumed by all involved that when HDTV becomes a serious threat to the film industry, 65mm film will once again be commonplace for production.

Many significant quality advantages exist for film over video in the areas of flexibility and durability. When filming dramatic subjects under adverse weather conditions, film possesses numerous advantages, due to its mechanical rather than electronic basis. Quality control is accomplished during the manufacturing process in film; in video, it is done on the set, at the time of production. This makes video far more vulnerable than film to misadjustments and daily set-up variations.

Despite its tinseltown, fantasyland image, motion-picture production is very much the art of the possible, the art of the practical. Although the motion-picture industry has the reputation in the popular media of being financially irresponsible, it is actually fiscally and technologically conservative. The upright moviola in the editing room, for example, is a classic mechanical contraption that exists today almost unchanged in style or function from the 1930s version. It is a fixture in the Hollywood editing suites. Although it is known to consume film and tear sprocket holes with alarming regularity, and although the improved horizontal flatbed-type editors have existed since the mid-1960s, the changeover is slow. My prediction is that at least in the Hollywood area, upright moviola editing machines will exist and be used daily to edit film until after the turn of the century.

This conservative streak does not mean that the motion-picture industry is without its pioneers. There are certainly those within the feature production community with a considerable taste and daring for innovation. However, this is in an industry where computers are being used to keep track of script pages shot per day and camera set-ups completed per hour, with daily reports sent to accountants and producers. When a director's job security is based on a computer projection of a project's completion date extrapolated from these daily reports, they are understandably leery of technological innovations on the set. The rewards for being the first director or producer to utilize new technology are media attention and notoriety for the innovators. The penalty for experimenting with new technology on a production, however, can be shooting delays and the most expensive kind of research and development—the type that goes on while the cast and crew stand around waiting.

What about this crew? One frequently hears that they are cheaper, or that less of them are required, with HDTV.

Crew cost comparisons between film and video production are quite difficult to make. This is because the price a crew will demand depends on the quality level of production and not on the technology used to record it. A crew hired to do an important dramatic film will be selected based on their experience and reputation; if they are in demand, their price will be high.

Producing such a project on video will not alter this fact. In the case of current videotape editing, we are even seeing that due to the complexity of the editing systems and their frequent obsolescence, trained editors are at a premium and may charge any price they wish.

An argument one frequently hears from the advocates of HDTV is that smaller crew sizes will be required. There is no evidence that this is true. Film camera crews consist of a camera operator, an assistant cameraman (to pull focus), and a dolly grip (to move the camera dolly). On three-camera TV comedy shows, it is true that one person is used per camera, and that this person operates, focuses, and moves the camera. This may reduce the crew by two people per camera compared to film; however, this method would not be feasible for a quality dramatic production. Even then, additional people to act as camera-control engineers and videotape operators are still needed. An even larger crew would be required for a serious dramatic feature produced in HDTV.

A comparison of union rate cards may show slightly lower rates for video than for film at the present time, in some job descriptions. Specialty jobs involving electronics, however, are very well paid. Practical experience indicates that in video shows with good production values, crews demand equal rates as in film.

Almost everyone—film makers and video engineers alike—seems to believe that video production is cheaper than film production. Many people who work in video production have promulgated this myth, and the irony is that they are constantly having to prove that it is true. When the client and the producer decide to shoot a commercial on videotape because its supposed to be cheaper, they then pay the crew less, use a smaller crew with less lighting and grip equipment, and have a tighter shooting schedule and a lower shooting ratio—all because video is supposed to be cheaper. The result, of course, is simply an inferior product. The truth is that "good" video is expensive—just as good film is expensive.

First of all, video equipment is expensive. Even if the cost of the camera is comparable to the cost of a film camera, once the recorder and the monitors are added, the cost of the video equipment is sure to exceed that of a film camera. And rental fees on video equipment tend to be higher because of the shorter life expectancy of video equipment. Video equipment is less durable than film equipment and also becomes obsolete more quickly. As a result it is necessary to charge higher rental prices in relation to the capital investment in order to have the equipment pay for itself before it becomes obsolete or wears out.

The claim is also made that video is cheaper because the tape can be recycled, but it is very rare that any producer will erase a tape in order to record new material on it except in off-line editing.

What would be desirable in a practical HDTV production system? A key point to understand when contrasting broadcast video production and motion-picture production is that, in broadcast production, the use of the equipment or facilities is one of the most expensive aspects of production, whereas in motion-pictures it is the least. In other words, you can't have Paul Newman standing around waiting for an HDTV camera to be adjusted and registered.

Today's HDTV cameras, while quite small, are connected by a cable to an electronics package the size of a telephone booth. Of this electronics package, perhaps one-sixth is camera signal processing, one-third is digital image enhancement, one-third is camera adjustment circuits and computer, and the remaining one-sixth is power supply. Removing the camera adjustment circuits and alignment computer during production would allow the reduction of power supply circuits. Improving the optical system and lenses might allow reduction of the image-enhancement circuit's volume. These changes may allow the camera, in time, to be battery powerable—which would be essential to its production usefulness.

The number of scanning lines in a video system is related closely to its potential resolution. The NHK theory of sufficient resolution for a high-definition television image (from Dr. Fujio's psychophysical research at the NHK Research Labs) said that, based on screen size and viewing distance, a maximum resolution could be achieved for a given size of picture such that increasing the resolution further would give an audience no increased experience of sharpness. Without going into excessive detail, suffice it to say that the NHK's own research indicated 1,600 lines was that figure for HDTV. For technical and political reasons a 1,600-line HDTV standard was never attempted. It is important to note, however, that when motion-picture special effects are generated through the use of computers (in non–real time), no line rate less than 4,000 lines is ever used. Several pictures, including *Tron*, *The Last Star Fighter*, and *2010*, have been made using computer-generated graphic images. I have spoken to the people responsible for these effects, and all of them stated they would not attempt motion-picture special-effects graphics with 1,125 lines.

In engineering circles, resolution is the first item on most lists of priorities where HDTV or other advanced imaging systems are concerned. Many engineers are perhaps overly defensive about matters of resolution in discussions of video's relationship to film. Many cinematographers feel it is not the superiority of film over video, in terms of absolute sharpness, which is the problem. A relatively small area of the picture, in any given frame, is critically sharp. The camera's depth of field limits critical focus to a relatively small area of the picture. The resolution characteristic of film, which is most difficult to imitate with electronic cameras, is the gradual transition from sharp areas of the image to soft-focus areas of the image. Due to the optical nature of this transition in film, it seems natural to our eyes, compared to the image-enhanced electronic appearance of video contours. This same transition in video, from an in-focus enhanced portion of the image to an out-of-focus and thus unenhanced portion, is most objectionable to film people.

When discussing resolution in motion-picture film, practical considerations must be kept in mind. Sufficient motion-picture sharpness for professional theatrical applications can be defined as follows: that amount of resolution that allows the cinematographer to put diffusion or nets in front of the lens, and deliberately toss away a large portion of the available resolution. That also allows the assistant, working with high-speed lenses, to miss his or her critical focus once in a while and still provide a result that can be screened to a large audience, on a projection system that is rarely in focus, without the audience stamping their feet and complaining. This implies that motion-picture resolution requires a reserve of sharpness to handle a certain amount of accidental or intentional abuse in the hands of cinematographers. The current HDTV standard does not provide this level of resolution.

A major concern in any relationship between film and video is frame rate. There are four principal considerations in setting the frame rate for an HDTV system: flicker, reproduction of motion, noise reduction, and bandwidth requirements. The higher the frame rate, the greater the bandwidth requirement of the systems, all other things being equal. The object is to choose as low a frame rate as possible without compromising the image quality, but frame rates as high as 80 frames per second have been discussed.

The threshold for visual perception of flicker is about 40 frames per second. Film projectors eliminate flicker by using double- or triple-bladed shutters, so that each frame is projected two or three times. With a repetition rate of 48 or 72 times per second there is no perceptible image flicker. Current broadcast television standards eliminate flicker by means of the interlace scanning of the image (lines are scanned alternately, on an even, odd, even, odd basis). It is the field rate rather than the frame rate that determines whether there is any perceptible flicker. NTSC has a field rate of 60 fields per second, which is comfortably above the threshold. With PAL and SECAM, the field rate is 50 per second, and a degree of flicker can be observed with exceptionally bright images, especially for people who are used to watching 60-field-per-second NTSC. Screen brightness of PAL receivers is generally held within limits that prevent noticeable flicker.

One of the considerations in designing an HDTV system is the question of whether it needs to be compatible with film frame rates. Early in the American HDTV standard-setting group study, the debate of 24 versus 60 versus 50 frame rates led to a study of frame conversions. A computer was programmed to simulate frame conversions and to search for a "magic" number: a frame rate that would convert into 24, 60, and 50 frames without compromises. Of course, none was found. It was found that all of the current frame conversion methods (such as 24 to 60 frames with a method called 3:2 pulldown) resulted in unacceptable image compromises when the resolution was improved to HDTV levels. Since one form of application envisioned for HDTV (especially in its initial implementation) is the production of movies to be transferred to film for conventional distribution, it may be advantageous to adopt a frame rate of 24 frames per second, or one that is divisible by 24 frames (for easy conversion).

Before leaving the subject of frame rates, it should be pointed out that while American broadcasters are opposed to a 50Hz frame rate for the same reason that European broadcasters are opposed to a 60Hz frame rate, American film makers may find 50 Hz more appealing than 60 Hz. This is because 24-frame film doesn't convert as well to 60Hz video as 25-frame film converts to 50Hz video. The difference between 24- and 25-frame film is of no great consequence.

There is a move afoot to convert the American (and in fact the world) film standard to 30 frames per second. This move is heartily encouraged by the advocates of the NHK system. It is quite unlikely that the film industry will ever really change to 30 frames per second, because this would destabilize the production industry's only true worldwide standard. Thirty-frame film production is becoming popular in the U.S. for productions destined *only* for video release. But despite a slight quality improvement in 30-frame film on projection screens, most film makers do not want to launch a new and unconvertible film standard.

Conversion to 30-frame-per-second film also creates financial and/or quality problems. If the film is shot in the standard frame size, increasing the frame rate from 24 to 30 means that an additional 25 percent of film is required for shooting. Those extra six frames per second can dramatically effect the cost of raw stock used in any production. The other option for increased frame rate is to decrease the exposure size of each frame to allow 30 frames to be shot on the same amount of film as 24 frames. In this option, there is obvious detriment to the image quality, as there is less film area exposed for each frame. Neither option is attractive to the majority of film makers.

As difficult as frame rate conversions can be for the reproduction of film motion on the screen, interlace scanning presents a bigger problem. Interlace scanning does solve two problems, which makes it ideal for *broadcast* video: It conserves bandwidth by allowing lower frame rates, without resulting in large-area picture flicker, and it allows a smaller number of scan lines to present the sharpness equivalent to twice the number of scan lines, again without the bandwidth penalty associated with doubling the scan lines.

The best method of eliminating large-area flicker if film transfer is involved is through the used of noninterlaced or progressive scanning. Progressive scanning, however, requires high frame rates to eliminate the resulting large-area flicker. High frame rates, in turn, require excessive bandwidth for broadcast and signal processing. One proposed solution to this spiral of complexity is to broadcast an interlaced signal and, through the used of a framestore, convert the signal at the receiver end to a progressively scanned signal for display.

These solutions, however, have no value when film transfer quality is paramount, in that interlaced scanning can result in motion aliasing, the effect of jagged edges produced when an object has moved during the time between the scanning of the first field of a frame and the second. The net effect is a loss of definition since the jagged edge is perceived as a blurred edge by the viewer. Higher frame rates help reduce this effect by reducing the time between the scanning passes and therefore reducing the difference between the position of the edge in one field and its position in the next field.

Frame rate and interlace problems make the NHK

standard particularly unappealing to film people. It is definitely true that motion artifacts are unacceptable to film makers, particularly since some applications may require double conversions (film to tape, then back to film) with the result of compounding the motion-portrayal problems.

It also appears that sequential-scan images of lower resolution can appear superior to an interlace-scan picture of higher resolution. The David Sarnoff Labs has demonstrated a black-and-white camera test system capable of producing a 750-line sequential-scan picture. In this demonstration, it appeared to have superior resolution than a 1,125-line black-and-white interlaced picture.

Overall sensitivity can be more important to an HDTV camera system than resolution. This is because sensitivity translates directly into production costs, whereas resolution does not. To realize the impact of increased sensitivity on a production, one must study what has occurred in film production with the advent of high-speed motion-picture film stock.

Sensitivity in a motion-picture stock, or indeed in a video production camera, directly translates into labor-hours required to light a set. Other hidden costs of low sensitivity are: electrical power consumption, caused by increased lighting and air-conditioning needs on the stage; fatigue factor on the part of actors; expensive production delays caused by the need to reapply make-up that has faded from heat and perspiration; as well as the length of time required to set up and strike a location of heavyweight compared to lightweight lighting instruments. It has certainly not been lost on the advocates of motion-picture technology that high sensitivity in relation to signal-to-noise ratio[2] figures is currently a major advantage of motion-picture stock over video technology. This advantage will increase rather than decrease as video technology progresses farther in the direction of high-definition video systems.

The ability to reproduce extremes of contrast is a critical requirement for any film or video camera. It is a difficult variable to quantify. In practical terms, a camera's ability to reproduce images of wide scene brightness range often is of primary importance to its photographic usefulness. Since lighting contrast ranges occurring in nature exceed the ability of both film and video to reproduce adequately, in professional photographic situations it is usually necessary to reduce the existing contrast range through the use of fill lighting. Put simplistically, the narrower the dynamic range that can be safely reproduced by a camera, the more fill lighting is required to reproduce a scene. Dynamic range limitations that do not occur in film cameras have always occurred in video cameras. In current state-of-the-art equipment, the video camera is at a disadvantage over a corresponding film camera.

While gamma compression devices[3] on current electronic cinematography cameras are incapable of increasing the camera's actual photographic latitude, they nevertheless simulate a filmic appearance in the reproduction of high contrast. An HDTV camera without gamma compression circuits would definitely be at a serious disadvantage. Current HDTV cameras do not have these circuits. The ideal, of course, would be a high-definition production camera with dynamic range beyond what is currently available, approximating or surpassing that of film. It is conceivable that charge-coupled device (CCD) technology may make this possible although it would, of course, be necessary to overcome first some of the CCD's other technical limitations.

Cinematographers will certainly identify themselves as those who communicate by making visual images with light. Video camerapeople tend to think of themselves as technicians rather than photographers.

This division between trust in the craftspeople and trust in the technological equipment exists nowhere as prominently as in the debate between film and video technologies. If you walk into a production company as a customer, the film production company's manager will show you his company's reel. The video production company manager, however, will give you a tour of the equipment. It is as if the video client is reassured by the company's technical capabilities and has little interest in its creative capabilities.

There is a concept in motion-picture camera design, that the highest state of the art is to produce a camera that is functionally transparent. The Panaflex film camera, for example, has been redesigned hundreds of times since its invention in 1972 but no one ever moved the on/off switch or the position of the speed control. In motion pictures, the best camera is one that a cameraperson looks *through*, not *at*...a camera that is so familiar that it becomes an extension of the cameraperson.

Since a camera is a tool to control light reproduction and contrasts, it is the image and photographic manipulations that are important. Cinematographers should feel more daring with video because of the immediate feedback it gives them. But in order to utilize that freedom, they have to be comfortable with the medium and the equipment. And, the camera must be designed to manipulate images, not just reproduce textbook ideal video images. Current video cameras, however, are designed around the philosophy of maintaining (by automatic iris if necessary) 100 IRE unit (standard-level) pictures. They were designed to meet the needs of broadcast engineers. Although video holds up the promise of responding to the cameraperson's creativity more immediately than film because of the instantaneous electronic nature of it—one should be able to get a photographic idea and realize it more immediately— cinematographers are still frustrated with this medium. Because of the relatively low priority that creative photographic control is given during the design of current video equipment, video production becomes more clumsy instead of less clumsy than the film production process, where you send the film off the set and view it the next day.

One of the main problems with the currently available video cameras is that designers don't expect the cameraperson to learn anything in order to use them. In other words, they don't expect the user of a broadcast production camera to be more of an expert than the user of a low-end industrial camera. They essentially give the professional cameraperson the same auto-white, auto-iris, and auto-registration as they give the amateur. These automatic functions are even present (and especially out of place) on the HDTV cameras.

It would be better to have the equipment be more versatile. Versatility does not mean more automatic functions, it means fewer automatic functions. It means more demands placed on the user while also giving him or her more versatility. The fact is, people who operate professional equipment are knowledgeable enough to make decisions and control functions of the camera according to their own intelligence and their own creativity.

## Endnotes

1. Reprinted with the permission of Richard K. Shafer, Eastman Kodak Company.
2. The signal-to-noise ratio (SNR) is a standard measurement of the amount of signal (desirable) compared to the amount of noise (undesirable) on a television image. Higher SNRs are preferable to lower ones.
3. A gamma compression device is a circuit that distorts the grey scale reproduction of video to make it more perfectly match the tonal reproduction of film.

# 23

## Producing in
## High Definition, Using the
## European 1250/50 System

*Michel Oudin*
*VP, Engineering and Development,*
*Société Française de Production*
*and Benedicte Delessalle*

Since the introduction of 1250/50 high-definition production equipment at the International Broadcasting Convention in Brighton, England, in 1988, about 35 European productions of all kinds have been produced (as of October 1989) in European HD. Among these productions were "Un Bel di Vedremo" by RAI Italia, "From Mozart with Love" performed by the Salzburg Marionetten Theater and produced in HD by ORF in Austria, "Zikus Knie" and "Gorbachev's visit in Bonn" by Germany's NDR, "The England Soccer Cup Final" and the "Wimbledon Tennis Championship" by the BBC, and "Gulenkian Ballet" by RTP in Portugal.

For its part, France, with SFP (Société Française de Production), covered the event called "The Night of the Bicentennial of the Revolution," which took place on the Champs Elysées. This event was presented in high definition to the president of the French Republic and 35 other heads of state who were present in France for the occasion.

SFP has also produced a 17-minute fictional piece entitled "1250, Any Better Offer?" ("1250 Qui Dit Mieux?"), which demonstrated what we believe to be the benefits of the European 1250-line production system.

The shooting of these programs allowed us to

experiment with the 1250/50 production tools and to draw conclusions on the organization of an HD production as well as giving us the opportunity to acquire the know-how necessary for such productions.

The production of "1250, Any Better Offer?" was targeted to illustrate the progressive and compatible development of this European system for television as well as cinema production; to demonstrate the complete range of available production, distribution, and transmission equipment; to show the reliability and performance of the European HD system through the production of high-quality images; and to demonstrate the potential of high definition in the creation of special effects.

In addition, the film was produced to convey the spirit of EUREKA, the European consortium for 1250/50 HDTV, and to show the expression of the European culture in programming, with references to cinema and with a touch of humor, even in a serious presentation about European 1250/50 HD.

The story of "1250 Qui Dit Mieux?" involves a crew that is shooting superb but classical HD images of the Chateau de Chantilly. The park gardener seems skeptical. He interrupts the shooting and asks questions about the different vans and equipment parked on the lawns. The narrator takes him on an amazing tour, where he meets Charles Chaplin and turn-of-the-century film maker Georges Melies on an HD telecine (film-to-tape transfer equipment), and Marlene Dietrich and Fellini's musicians at a special-effects mixing board.

The film is based on the extensive use of special effects, where the characters mingle with HD equipment. It was shot with means equivalent to those of a first-class fictional production, including sets, costumes, and make-up, and with set construction representing an important part of the budget, mainly due to the extensive use of special effects. The production was shot in three weeks and the final cut includes 60 scenes.

It is commonly said that HDTV is neither cinema nor video but "something else." We would like to agree that high-definition production is a new media. It is video-style production with film quality.

An HDTV camera is essentially identical to a standard 525- or 625-line camera in size and weight, except for the size of its lenses. Indeed, the lenses must cover a broader image in a larger format, while at the same time maintaining maximum apertures adapted to the sensitivity of the tubes.

It is therefore necessary to design two types of cameras: a "live" camera to cover various types of events, like sports, where a vast range of optical focals is needed and a "fiction"-type camera, or a "single camera," with either fixed focals or smaller zooms adapted to specific shooting needs. Monitors are an important part of any production, since they allow immediate control of the image. These monitors must be of the biggest possible screen size and optimum reliability. It is also very useful to complete the screening by using a video projection system that will reveal defects in the details of the images.

All equipment other than cameras and monitors, such as videotape recorders, camera control units (CCUs), sound recorders, video mixers (switchers), etc., are gathered in a van, which is connected to the set by a net of cables. The van gives a certain mobility to the equipment, so it can be used for shooting on location or in a studio. However, we are still far from the versatility of film production, where the camera is completely autonomous.

The sensitivity of the HD camera is almost equal to that of a 525 or 625 camera. It is in the areas of contrast and colorimetry that HD shows its superiority over standard video images. They seem better because, in HD, the sharpness of the image highlights the image's other characteristics. Colorimetry is becoming an artistic tool; thus HD's improved colorimetry is greatly appreciated by the director and the technical crew.

An HD shoot is often compared to a film shoot since both use mono, or single, cameras. Monocamera shooting concentrates on a single axis at a time, thereby placing emphasis on direction and photography. HDTV also allows for multicamera shooting of special events, however. For instance, the SFP coverage of the Bicentennial Night used three cameras.

The fundamental characteristic of a video shoot is the immediate control of takes by the crew. In some cases, this control turns out to be a waste of time, as it generates excessive discussions on adjustments and settings and leads to retakes. At the SFP, the technical crew that shoots in HDTV is mostly formed of video technicians, who we feel are better armed than film crews to deal with the technical aspects of the set.

For post-production, we transfer the images to low definition for off-line editing. This process enables us to avoid using an HD videotape recorder for the off-line editing and allows us to work from the original master when producing the final edited master. However, the process also requires extreme precision during the off-line process in order to create a computerized edit list that can be used when conforming to the master during the on-line edit. This need for precision emphasizes the importance to be given to synchronization signals and time codes while recording. As a security measure, the original material and off-line editing codes are visually inserted onto the frame, thus allowing us to manually reconstruct the edit list, if necessary.

The only artistic difference between standard video editing and HD editing is that fully perceiving all the detail in an HD shot on a wide screen takes more time than a shot in low resolution. Therefore, we have found that HD productions have a slower rhythm than standard video.

Our soundtrack for "1250, Any Better Offers?" was originally recorded in stereo analog mode at the SFP. We used a three-track stereo system to record it: a pair of mics forming a 110-degree angle and a directional mic in the center for the later mixing of close-ups. A special sound mixing was done for later broadcast and screening in "surround" sound. The special mixing was based on a four-track sound played on five speakers: three in the front and two in the rear.

For special effects, we used an Ultimatte compositing system and a switcher adapted for 1250/50 HD. The special-effects process was not altogether different from the one used in standard video. But because we were working with analog HD, care had to be taken to record all of the live effects at the first generation, to avoid the degradation of the image that would occur in multiple generations.

The director's approach to the production was to increase the feeling of reality by creating interactions between the actors and the sets. Thus, we had to build numerous blue sets and be careful while shooting to conform our blue sets (for matting) with the miniatures that were built for the matte effects and set backgrounds and were eventually composited with the live action. The match between the miniature and full-set perspectives, the ability of the Ultimatte to render projected shadows, and the quality of the HD image combined to make the special effects quite spectacular. On purely video aspects, such as real-looking special effects and color and contrast rendering, the HD image acquires an artistic dimension that is lacking in standard video.

The director of photography of "1250 Qui Dit Mieux?" said that what best defines the HD image is its high-fidelity capability to render the smallest details. Therefore, the largest possible screen should be used for viewing dailies, so that all the detail can be seen fully. The 16-by-9 screen format corresponds to an aspect ratio of 1.77:1 and is almost compatible with both the 1.66:1 film format used in Europe and the 1.85:1 format used in the United States. The wider screen requires a different concept of direction than standard video. It is our recommendation that the cameraperson be trained in shooting 35mm film, which has a similar aspect ratio.

However, the widescreen format may not be the most appropriate for certain television programs, such as talk shows or news. The 4-by-3 format of standard television is well suited for "talking heads," while the 16-by-9 format is more suitable for a two-way conversation. The 16-by-9 screen size also requires that a greater importance be placed on the background, however.

High-definition production requires very precise shooting techniques, more akin to film than video. In HD, the depth of field is shorter than in standard video and focus errors are therefore more noticeable. For this reason, the precision of film methods should be applied to HD production. The focus puller in HD must measure the distance between the subject and the camera at a different stages of the take, as in film. When the subject moves and thus the distance changes, the focus puller will then be able to focus correctly without checking the focus through the camera lens.

One difference between film and HD is that in HD the focus and depth of field of the final image can be immediately checked on the monitor; both the focus and depth of field can also be accomplished by remote control, often from the production equipment van. By using the monitor and having on-site control of the images, depth of field becomes an artistic tool in HD as much as, if not more than, in 35mm film production.

Because of the details that can be seen in HD, the sets, costumes, and make-up can be seen in greater detail and require better design and construction than in standard video, which can result in costs as much as twice as high for HD. Special effects also require special attention and greater resources. To this end, the SFP structure, which integrates set design and construction, photographic and graphic workshops, video and film facilities, and technical and creative teams, thus offering all the elements required for special effects.

During the shooting of "1250 Qui Dit Mieux?" we used various references to the cinema and numerous motion-picture clips. Due to the compatibility between 24/25-frame-per-second film and the 50Hz European HD system, film clips were easily used as shots or background elements in the HD production. We expect that in time, with the development of HD-to-film transfer equipment, video special effects will also be able to be inserted easily into motion pictures.

The 50Hz frequency compatibility creates an immediate and easy link between European HD and film. The further development of the 1250/50 production standard will provide an even better compatibility with motion pictures than currently exists, and will tremendously improve the quality of HD images transferred to 35mm film.

The ability to use video tools in film production for viewing, editing, and special effects, and the ability to easily transfer from HD to film (and vice versa) will mean a complementary instead of competitive relationship between the two media and cultures. Additionally, the compatibility between European HD and film will facilitate the distribution in HD of feature films in their original quality.

European countries are currently producing HD films and are jointly establishing coproduction ventures in 1250 high definition. This will enable broadcasters and producers to expand their experiments in HD and to acquire the necessary know-how. And it will prepare them for 1992, which is not only the year of increased economic ties in the European Community, but will also be the beginning of experimental broadcasts of HD programs. And, in 1995, European satellite and cable broadcasting of HD are scheduled to begin.

The initial concern of the EUREKA program was to concentrate on research and development of equipment. Eureka is now incorporating broadcasters, producers, and manufacturers in its efforts. Europe will succeed in uniting its program resources with its potential in technological innovation. Then we will be able to benefit from a tight link between the most artistic media, cinema, and the most advanced electronic image technology, high definition.

As a consequence, high definition will become the medium for first-rate programs, both fictional shows and live special events. And with an audience of 350 million, Europe has the resources to have independent production and distribution networks for this type of high-quality program.

*Editor's note:* This chapter was excerpted and adapted from a technical paper presented by Mr. Oudin at the 130th SMPTE Technical Conference in Los Angeles, October, 1989.

# 24

# HDTV: Toward a New
# Visual Vocabulary

*Stuart Samuels*
*Vice President,*
*Zbig Vision Ltd.*

*H*DTV is the most revolutionary imaging system yet created by human society. It will usher in the next phase of picture making—capturing the contemporary technological mantle previously claimed by drawing, painting, photography, 35mm film, and traditional video systems. First, there was painting—reproductions of reality. Then there was photography. Film made pictures move. Video gave us live transmission of images. Now, HDTV will enable us to create a new kind of visual vocabulary. If the film or video maker can think of a visual image, he or she now has the tool to actualize that image on tape and show it to society. HDTV as a production medium is more versatile and more extensive than film in its ability to create complex images. It is a technology that brings creative thinking much closer to creative production.

HDTV combines the image quality of 35mm film with all the technological advantages of electronics, especially those associated with computers. HDTV is the first production system to challenge the acknowledged supremacy of 35mm film.

HDTV brings visualization into the computer age. Film is a 19th-century technology founded on chemistry and mechanics; HDTV is electronic. It links image making with state-of-the-art technology, combining all the advantages of live transmission and computer programming.

The most obvious aspect of HDTV is that it presents a better-quality picture than that currently being shown on our present TV screen. HDTV, no matter what line resolution or frame rate, is the closest image-making system we have to everyday reality—35mm film has grain, and reality doesn't exist inside a grainy visual environment. Traditional NTSC or PAL video systems, with their limited color rendition and depth of field, are also clearly removed from what we see around us with our own eyes. Therefore HDTV brings us closer to perceived reality than any other imaging system.

The HDTV aspect ratio (16:9) is also closer to the way our vision works than today's television aspect ratio. The idea of a box-like image system—like our current TV—does not correspond to the way our eyes take in images. The images we would see on an HDTV screen would more closely correspond to the way we see things in nature.

These improvements in picture quality and aspect ratio will change the way we consume sports, music, entertainment, and movies inside the home environment. The baseball or football game on HDTV will look more like the game seen in person. The ability to have a shot of a symphony orchestra in medium close-up and still encompass the complete orchestra will position us at home as closely as possible to our real-life experiences of classical music inside a symphony hall. The true stereo quality of HDTV sound will rival CD recordings, so listening to a rock concert on large-screen HDTV will give the viewer a concert-like experience inside his or her own home.

The current negative attitude toward the "video" look compared to the look of film—so common in the advertising and film-making worlds—will lose the power it holds over creativity, since feature films or miniseries made and shown in HDTV—without grain—will narrow the creative distance between reality and the illusion of reality.

But picture quality is not necessarily the most

important factor in gauging the impact of images on popular culture. Perhaps the most significant image—moving or still—of the 20th century is a badly out-of-focus, degraded shot of a man in a space suit taking the first steps on the moon.

Thus, although HDTV's improved picture quality will have a significant impact on our future, it is not the picture quality that is the most revolutionary thing about HDTV. It's the *kinds* of pictures you can produce with this imaging tool that will deliver HDTV's most revolutionary impact on our culture. The kinds of images HDTV as a production tool can produce will usher in a new kind of visual vocabulary more suitable for the last decade of this century and more relevant to the image needs of the 21st century.

The most culturally revolutionary aspect of HDTV will be in its production capabilities rather than in its transmission and exhibition qualities. HDTV has the ability to transform the way we look at our contemporary world and at the future. HDTV can bring magic back to the big and little screen. It can transform our visual vocabulary the way that Cubism and Surrealism altered our view of static pictures, the way Melies and Lumière changed moving image, the way films like *Star Wars* with its realistic depiction of a future reality altered the way we looked at our world and ourselves.

The quality of HDTV that will change the kinds of films and videos we make is its ability to composite images by seamlessly matting one image inside a previously shot background or scene. Matting is the superimposition of one image on another. While 35mm film is limited by expensive optical special effects, HDTV, because of its seamless matting process, offers possibilities for radical new kinds of imaging. Film as an image technology has been largely explored. And, because of its mechanical and chemical foundation, the composition of more than four separate visual optical elements with 35mm, or even 70mm, film produces unacceptable results.

With HDTV production, on the other hand, you can multiply components—visual elements—and separately control all elements of the shooting. This method of production will lead to the development of new special visual effects at a substantially reduced cost, and with a greater degree of control over the picture.

HDTV production, when linked to a blue-screen process called Ultimatting, can composite layer upon layer of separate images inside each individual take. By creating images in this way, a director can have greater control over the exact image he or she is trying to create. The kind of image compositing is closer to the style of 24-track audio recording methods than to traditional film or video making. If a director has the ability to combine up to 15 or 20—or with digital HDTV, unlimited—visual elements to create a scene, the kinds of scenes shot will change, as will the nature of the stories filmed.

The current special-effects economics of 35mm film making relegates effects film making to the highest budget category, and therefore requires sure-fire hits and popular themes to lower the financial risk. Special-effects film making for medium or lower budget films has resulted in a reliance of horror special effects, which require more make-up effects than optical tricks, or obviously fake or campy send-ups of special-effects film making, which don't demand a very high level of realism.

The ability to combine multiple generations of images electronically, matting each one in a way that, with the proper perspective and understanding of geometry, will look as if the actors realistically inhabit the environment in which they are placed, will revolutionize the film and video industry. Scripts long gathering dust on studio and agents' shelves will be looked at in the new light of affordable special effects. Films rejected because the effects necessary to deliver the film were going to be too expensive can now be assessed on their subject matter rather than their special effects requirements.

HDTV can create a new kind of visual vocabulary, so dense with possible detail as to rival the look of a Bosch painting, the multilayered synthesized audio creations of new age music, or the surreal content of our wildest dreams, our most horrifying nightmares, and our most far-out and far-off fantasies.

One of the casualties of our post-modern culture has been the absence of any significant Utopian vision. As we enter into the last decade of this century, our intellectuals will look into the future, as is customary, to conjure up a vision of the next century. With the revolutionary ability of HDTV production to create multilayered imaging in a realistic perspective, film makers (and lower-budget ones at that) will turn their considerable visual talent to creating big- and little-screen images of the future.

As Zbig Rybczynski has said:

> "You can create with this new technology something we have never seen, never experienced. It gives us the possibility of a new reality the way Disney or Chaplin or *Star Wars* did. It can make our world, our imagination, richer."

Because of its linkage with the computer and all other computer-based imaging technology (graphic systems, motion control, digital video-effects generators, and all future devices the electronics makers have up their technological sleeves), all elements of the HDTV picture can be precisely and exactly—down to the exact field—controlled by the creative imagination of the director with the aid of his engineers and editors. The ability of the director to do live editing during the production (that is, compositing live with the camera and video machines) will revolutionize the way films and videos are made. Traditionally a film is completed many months after it is shot. In fact, most actors have little sense of continuity or finished story line until, in some cases, a year after the shooting. This method gives the most important role to the director inside the editing room, going over the film, frame by frame, cutting together a vision that often does not correspond to what has been shot or what is necessary, but that is impossible to reshoot.

In a similar way, current video technology, with all its special-effects toys, puts greater emphasis on "fixing" things, or creating the "look" of a work, in the post-production process.

Imagine a different situation, one that corresponds more closely to the state of current HDTV production technology. By using HDTV you could prepare all the necessary elements and store them on separate tapes. Then in a blue-screen studio you could seamlessly composite into these elements the live actors. This is a completely new situation for directors, actors, and crew and it encourages production spontaneity and innovation.

Whereas the final shape of a film production traditionally takes place long after the shooting, now imagine a film- or image-making process that works as follows:

1. A director and writer join forces before a script is done to make a film that takes advantage of the best aspects of the technology while working at the most creative level of imagination.
2. Once a combined writing and special effects script has been prepared, the period of extensive preproduction begins.
3. The background scenes or model building required to create the visual look of the film are shot or created inside an HD effects studio, without actors.
4. With an HD camera, a few video machines, a portable motion-control system, a graphics system, a small crew, and engineers, the director can create the visual look of the film—the background sets in which he or she will place actors and props inside the more controlled environment of a blue-screen studio.
5. The video editor and engineers prepare the edited backgrounds into a rough edit of a completed film. All special mattes required can be pre-prepared. Even an approximate sound track can be edited onto the background master.
6. With the backgrounds pre-edited, with the special matting effects completed, with the sound track added, it is now time for the actors and for the production itself. In a concentrated period of time, the director, with the actors and other assistants—lighting, make-up, costumes, props, engineers, graphic artist, motion control expert—all

acting as a creative group, will make the finished film in real time. In this way the collaborative nature of film making will not only refer to the number of people necessary to make a film, but to the creative collaborative nature of the film-making experience itself. Making films in HDTV will be a more totally creative and spontaneous process than in other media.

This approach to image making will bring changes to the way we make our movies and TV products. There will be a more scientific approach to image making, one that will rely more on exact measurement than on the director's or cinematographer's intuitive eye. Engineers will play a larger role in the production. Short complex takes will replace longer takes. The film maker will become an electronic painter. Inside the frame, compositing will replace image montaging. Actors can see at the moment they do a scene exactly how it will look, and also see the previous takes or subsequent take to gauge the proper emotional and acting needs of the scene. Film making will become a much more exacting rather than intuitive process. It will also become more spontaneous and hopefully more creative at the time of the production, rather than having the creativity relegated to the more solitary process of post-production.

The expected result will be the making of different kinds of films. This in turn will affect our values, our vision, our hopes and fears, in the same way that all image systems in the past have impacted our lives. HDTV offers us the possibility of revitalizing our culture by giving us a tool for a creative vision of the future based solely on our imagination. If you can imagine it, HDTV can create a realistic likeness of that vision.

APPLICATIONS: CURRENT AND FUTURE

# 25

# High-Definition Television and Medicine

*Robin J. Willcourt, M.D.*
*President,*
*The HD Pacific Company*

*T*he power of the picture is obvious to us all. From line drawings on a cave wall to the stunning videographics seen in special-effects movies, the human has always been compelled to communicate through pictures. Today, the startling pictures obtained by high-definition television (HDTV) elevate the power of this form of communication to an entirely new level. The detail that can be reproduced by the HDTV cameras and monitors boosts television's already awesome powers to communicate. There are many nonbroadcast areas that will benefit from this technology. Medicine is one of them.

Medicine is a highly visually oriented discipline. From the earliest days as a medical student grappling with the text and pictures of an anatomy book or suffering eyestrain through hours of peering into a microscope lens of cross-sections of organs, from the reading of an X-ray to the subjective impressions of a patient's appearance, the physician has primarily used the gift of sight to define and diagnose the mysteries of that discipline we call "medicine."

It is interesting to note that medicine is still taught as an apprenticeship, much as it has been over the centuries. Certainly, a great number of new tools have been added to the teaching process, such as interactive video simulations for case management and lifelike mannequins on which to perform CPR, but the basic sciences and clinical teaching are still largely taught through classroom lectures and bedside demonstration. This method, then, effectively determines the number of students who can be taught at any one time by the lecturer or clinical teacher. It is extremely important for the student to avail him or herself of the opportunities that apprenticeship teaching provides. It is also clear that the apprenticeship offers variety more than consistency, through personal interaction with an instructor in place of the impersonal textbook or computer interaction. Perhaps a sense of personal growth and development may be communicated by the rapport with the instructing physician, which is as much a part of becoming a good physician as medical knowledge.

The pathway to full medical licensure takes a student of medicine from the basic anatomy instruction to a full round of clinical teaching that has the student examining the patient from top to bottom. This didactic textbook and classroom teaching mixed with the precarious and personal intrusion into the patient's innermost mental and physical loci places an enormous burden on the physician-teacher, student, and patient alike. Yet, to date, no truly effective ancillary aid has appeared to improve the burden placed on all these participants.

It would seem that television should be one of the most utilized aids for the modern-day physician, given its capabilities of providing instantaneous visual information, of covering vast distances, and of allowing review and archiving of information. Television offers the promise of shifting the onus of disseminating information from the medical teacher to a piece of equipment capable of augmenting the visual aspects of teaching. This teaching can be achieved in a manner that is less intrusive and certainly less critical of the student's actions. With video, the student can move at a pace that is appropriate and comfortable as he or she explores and assimilates the never-ending volume of knowledge being offered.

However, as any inspection of a medical school campus or hospital complex will show, television has not lived up to its promise, which was to provide another set of eyes, to give another perspective, to see more clearly, and to communicate information unequivocally. The reason for this underutilization is that television, at its current low state of resolution, is neither a substitute for nor an effective ancillary aid to the eyes of the physician, who requires a visual medium that will not inject artifacts, degrade the information, or hinder his or her ability to practice the art and science of medicine in accordance with the public's high expectations.

Many of these concerns are being removed with the arrival of high-definition television. The resolution of the HDTV screen is at least double that of conventional television with about five times the actual discrete visual material on the screen. The wide screen and accurate color information provide much of the detail that needs to be conveyed to the student. High-definition television can augment the teacher's abilities and, in some settings, it can substitute for the teacher.

Anatomy is one of the basic clinical sciences that must be fully grasped by the student. This is an enormous task for most students since the information that must be absorbed is detailed and voluminous. Classic anatomy texts are often longer than 1,000 pages, with a lot of small print and highly detailed drawings. The language and nomenclature is foreign and the precise though complex manner in which the texts are written makes the task of absorbing and understanding the information one of the greatest chores of medical school. Yet, this is a subject that must be understood, as it is a base for all else that follows. Adding to the difficulty of grasping the volume of data that must be learned is the study of the cadaver, which is meant to clarify the relationships of the anatomical structures with respect to each other. Clearly, the student must be able to explore the real body in order to become familiar with the size, shape, and spatial relationships of the various organs. This study is accomplished with a further cost to the cerebrations of the student, who must now try to integrate the textbook data (text, photos, and diagrams) with a cadaver specimen.

The textbook information and the appearance of the cadaver bear little resemblance to each other and, what is worse, virtually none to the living human soon to be explored by the student.

This method of teaching anatomy has not changed significantly over the last couple of centuries, although there are examples of video instruction used as adjuncts to the conventional methods of teaching. Most video material is itself limited by its ability to render color, texture, and detail and so is not able to add significantly to the student's appreciation of the subject. Thus the role of current television has been simply to convey data. But even this goal has not been so easily achieved. In fact, there is a great deal of inefficiency in current television in trying to teach a subject that presents itself as a veritable chameleon, with the video images changing according to the factors of video production, including lighting, camera set-up, and even close-ups versus wide shots. This is fatiguing to the brain and only adds to the time it takes for the student to grasp all the details that need to be absorbed. This is one area in which HDTV will shine. It offers a real solution to the dilemma of integrating the sense and meaning of disparate images. A controlled flow of visual information, with the images looking like the actual appearance of the organs, allows the organ's structure and relationships to be fully examined through the clarity and reality of the HD picture.

Clearly, the student will have to deal with the cadaver, but this will no longer be the primary source of the visual input. The cadaver will only be one representation of reality, a means to understand the three-dimensional aspects of anatomical relationships and not the primary visual source upon which the student relies for his or her perception of reality. HDTV images can be built from a database that uses actual HD video of surgical procedures. The real-life anatomy, accompanied by colorful stylized 3D graphics of the organ systems, used in conjunction with an appropriately designed anatomy text, should shorten the time it takes the student to learn this material because he or she will relate it to the actual surgical examples encountered in clinical medicine.

Today, the student entering the operating room for the first time is confronted with a flood of new visual information that does not resemble anything previously encountered in the basic anatomy course. This is where the process of retraining the brain with new images while trying to recall the names that were previously attached to the standard images begins. By now both the names and images lie in the far recesses of the grey matter because both have been pigeonholed for some time; anatomy is learned well in advance of any exposure to surgery. These new sets of visual data must now be integrated with the old, along with names of the nerves, arteries, etc. It is not surprising that students and residents in training must spend many long hours trying to learn and maintain this vast amount of information.

I believe there should be an international effort to produce standard HDTV video texts of anatomy, physiology, histology (the study of the human organism under the microscope), and pathology (the study of the abnormalities of the human organism under the microscope). By producing these "texts," the best possible images from the world's vast libraries can be shared by students living in either the richest or the poorest countries. This international effort would allow information that is vital to all in the field of medicine to be applied without regard to race or social status, so that each physician, wherever he or she may live, is comparably educated.

Another use for HDTV in medicine would occur in the literally thousands of commercial pathology laboratories in the United States. These laboratories process tissue specimens, from simple urine tests to complex sections of organs, to help provide the treating physician with information about the patient's condition and effectiveness of treatment. These laboratories often comprise a central main laboratory and many satellites, which are the entry points for most patients in each geographic location. Here, simple tests such as blood smears and non-complex tissue processing take place. The materials are then sent to the central laboratory for reading and interpretation.

This process lends itself to the HDTV technology in

a couple of ways. Firstly, the preparation of the specimens in the peripheral laboratories means that the patients do not have to go to an inconvenient location. These specimens, as well as tissue specimens collected from surgery, could be processed in the peripheral laboratories in greater numbers than are now done, saving time and costs in the transport of the material to the central laboratories. Using HDTV scanning of the microscopic specimens, the images could be stored on videodisc, waiting for the pathologist to read them. The images could then be transmitted to the central laboratory when the pathologist is ready. By means of remote control, the pathologist could "scan" any portion of the image, make the appropriate diagnosis, and prepare a report. In some cases the pathologist would require some manipulation of the slides by technicians at the peripheral laboratories while in other instances the data would be insufficient for diagnostic purposes and the specimens would still have to be sent to the main laboratory. Still, a large amount of material could be handled without transporting it to the main laboratories, which would produce savings in time and cost to both the consumer and the laboratories. The second material benefit from using HDTV in this way is the ability to store these images for a long time and have the ability to archive and retrieve data in a very efficient manner from the storage devices, be they videodiscs or other storage media. This storage allows for easy comparison of specimens. It also allows the flexibility inherent in a system that permits recall of data at one location and transmittal to other locations to be reviewed by a larger number of people than is now possible.

Medicine does currently use television in one very specific area: medical imaging. The quality of the image notwithstanding, it has been of great convenience for the imaging centers to have their images captured and stored in a television format. It has always been understood that the quality of the images is not as good as the source material in many instances, but the ease with which the images can be recalled and displayed has overcome most of the obstacles to the use of television in this, the most visual of all the branches of medicine.

There is, of course, always the primary mode of image capture to fall back on when the quality of the TV image would not suffice. For example, while some X-ray images are stored in the TV format, the original films are sought if there is any diagnostic dilemma occurring by trying to interpret the low-resolution TV images.

To put this use of TV technology in perspective, it is important to understand the chaotic situation that actually exists in all medical imaging departments. In some medical centers, up to 500,000 separate X-rays may be taken in a year. This does not include CAT scan or MRI imaging, nor the use of moving pictures such as angiography. In all these cases there are dozens of separate images taken, which must all be captured and stored. Worse, none of these recording or display media use standard television formats, but totally different systems of recording that are specific to given medical video media, so an image stored in one system may not be viewable on other equipment. It makes the broadcast television NTSC/PAL/SECAM fiasco seem trivial by comparison. Ultrasound is another imaging service that adds to the problem of storage and display. These concerns have taxed many medical centers and diagnostic facilities that handle a high patient load, and the situation is worsening each year as the number of tests per head of population increases, so that storage space availability and the costs of the storage have become significant managerial factors for hospitals and medical facilities.

To add further to the chaos, let it be noted that each of these imaging services uses its own unique methods for the primary display of the image, quite separate from the means of storage. For example, X-rays are captured on film and displayed on a viewing box, a device consisting of fluorescent lights in an enclosure that has an opaque faceplate. The X-ray film is positioned on the faceplate, the light shines through the film, and the image can be viewed. This film is then returned to a paper envelope and is placed in a storage system, deep within the bowels of the medical center or diagnostic laboratory. Trying to retrieve that envelope is no mean feat, but showing the images to someone not in the viewing room is impossible. This scenario holds true for all the different imaging systems. The ultrasound video output has a different number of scanning lines than the MRI or CAT scanners and so the idea of storing them on a single medium has been impossible. High definition offers a valid solution to this problem, since the quality of the system meets all the demands posed by the highest quality source, X-rays. Given that HDTV will not visibly degrade the primary image source, it is then possible to design a system that will allow a common storage and display—the videodisc and high-definition monitor. Then, a common medical image format is needed to allow all the various imaged outputs to be transmitted in a compatible way.

To underline the importance of this, consider the typical scenario of a patient who has had chest pain with occasional dizziness and some numbness and tingling in the right hand. This patient is likely to be investigated for both a cardiac and a nervous system origin for the symptoms. Assuming that there is strong presumptive evidence for a cardiac etiology for the chest pain, that patient may have a cine-angiogram, where dye is used to show the coronary arteries. The movies are stored on conventional film, although the output is often put on VHS tape for review purposes. The patient will have an ultrasound of the neck arteries and either a CAT or MRI scan to detect any brain pathology. Each of these images is captured and displayed on different media, in different formats. Now imagine that the physician wishes to discuss this case at a medical conference in the hospital with colleagues. A different display device and play-back medium must be used for each of the tests. Clearly, the images cannot be sent anywhere for consultation, except by mail.

HDTV provides a ready answer to this pressing problem. The output from each of the imaging devices could be in the HD format, and sent to a central location where the images are stored. Since each is tagged with the patient's ID number, it is possible to recall all of these images at any time and review them on a standard display device. A conference on any patient becomes a whole lot simpler with this system, since all the data can be displayed in any sequence desired, but on the same display device. In addition, the images can be transmitted to remote locations and, to simplify things even further, if the patient suddenly appears at another facility in the middle of the night, it would be possible to call up all the video data and review it in moments, rather than hours, days, or not at all.

Additionally, the savings in storage space alone, despite its pedestrian nature, makes HD imaging a worthy goal, while the upholding of the Hippocratic ideal makes it exemplary.

One comment that always arises about the quality of the HDTV image when compared to film is that film is able to resolve around 4,000 lines, and the 1125/60 or 1250/50 systems can only resolve a little over 1,000 lines. This statement is usually uttered with the greatest concern by the makers of the X-ray films. It is clearly in their interest to take this position and, indeed, they are correct. But only so far as theory goes. In practice, it is unlikely that the average X-ray film has close to 1,000 lines of *resolved* detail because of the manner in which X-rays must pass through the body. In so doing, they are scattered by the tissues through which they pass. This leads to an obliteration of fine detail in areas where there is dense tissue, bone for example, and causes artifacts to appear in the final print. In addition, the real contrast ratio from any given print is always poor, even though X-ray film has the inherent capacity to render extraordinary detail with a good range of contrast. In reality, this is not achieved due to the limitations of the techniques of radiographic imaging. Subtracting further from the quality of the viewed image are the limitations of the viewing box, a device that scatters light throughout the film. It is a low-contrast source and as a whole degrades the image even further. To many people, a superb 525-line black-and-white image offers a superior viewing experience than X-ray film on a viewing box, and this is even more true for an HDTV image.

Some changes to the HDTV standard in terms of scanning could be made for medical applications, but even if this were not done, the current proposals using over 1,000 scan lines should meet the requirements of all but the most stringent critics, and they, by all accounts, should not be satisfied with any of the current video imaging systems.

The majority of residency training programs, like the basic sciences teaching described earlier, are based on the premise that one sees a procedure, does it, and then teaches it! This makes the most of a bad situation, namely that the availability of patients willing to undergo a surgical procedure at the hands of an inexperienced surgeon, no matter how well supervised, is shrinking and one must make the most of the experiences that arise. In a surgical residency, learning is accomplished by undergoing an apprenticeship with supervision by the patient's own doctor who performs any of the surgery that requires a level of skill greater than that of the resident. This process allows the resident to do whatever is within his or her level of training. By close and frequent observation of each surgical procedure and by participating at ever increasing levels of complexity, the resident finally is able to be certified as a specialist. This is really no different from learning to drive a car or fly a plane. Eventually, the trainee becomes a trainer. Still, the path to this glorious end is not so easy. There is the great complexity of the subject to contend with, not to mention the honing of physical and intellectual skills. This is the classic apprenticeship.

HDTV brings such clarity of information that it is now possible to produce videodiscs of each surgical procedure so that the student and resident can become familiar with each procedure before ever venturing into the operating room. This capability alone increases the value of the experience in the operating room, allowing the student to concentrate on the techniques of the surgical procedure, rather than spending the first few exposures to a new procedure simply trying to relate all the anatomy that was so poorly presented in the basic clinical science years. It is to be expected that no surgical procedure can be learned from a video presentation alone, but the capacity for interactive video learning for surgery is vastly improved by the use of HDTV. It is conceivable that many surgical training programs could be reduced by a year or so with an effective HDTV system integrated into a medical school curriculum from its beginning. The use of "juke-box" style videodisc arrangements, with each program accessible through an interactive computerized set-up, is a readily achievable goal and one that should be appealing to faculty and students alike.

The ability to tap into the vast amount of talented physicians whose surgical skills can be recorded and used around the world as benchmarks and in the development of standards argues strongly for this approach to the teaching of medicine, which is an international enterprise and not a regional one. Medicine is too often rationed, often because of lack of knowledge on the part of the providers. This is especially true in underdeveloped regions.

Live television has a place in the operating room in addition to the application already discussed. Many surgical procedures now are done with various forms of "scopes." By entering the body through a relatively small incision, the surgeon is able to inflict less trauma, the patient experiences less post-operative pain, and the healing process is enhanced. However, only the person using the "scope" can usually see what is going on, although attachments to the instruments have been made to allow one other observer to view the proceedings. Still, the power of television is such that it can allow a much larger audience in on the procedure and the image can be sent outside the operating room to other observers. By using HDTV, the images are clearer than today's television and can be displayed on a large screen without picture degradation, enlarging the viewing audience even more. Perhaps of more value is the fact that the surgeon now has a superb view and can operate with great confidence from the HDTV screen.

Remote diagnosis is also possible with the high quality of the HDTV image. In inaccessible parts of the world, an HDTV transmission from a remote "diagnostics" van would allow specialists at a major center to discuss the patient under the camera's eye with the patient's primary care giver, who would be in the van with the patient. In many instances, this form of teleconferencing would spare the patient the need to travel to the main medical center. Even with the relatively high costs of HDTV equipment, this form of medical intervention would be cost-effective and of great benefit to those who live in remote and poor areas of the world. Basic public health information could also be communicated to whole villages, using the technological infrastructure that would eventually be set up in these areas. As the reliability of the hardware increases and the costs fall, it would take little in the way of personnel and maintenance costs to keep a basic system intact.

By now it should be clear that high-definition television enhances the value of medical services. Its color purity, the ability to render texture and clarity with an almost three-dimensional appearance, and the wide screen can change the way medicine is taught and learned. The imaging and the pathology services can be greatly served by this medium. The current cost of HDTV might preclude some of the wide-ranging applications. But, if costs come down, as they are expected to do with increased manufacturing and demand for equipment, HDTV becomes an effective as well as cost-efficient medical tool. Remote diagnosis and the provision of services to the undeveloped areas of the world through the use of accurate images, linking the sophisticated with the have-nots, and bringing our halls of great medical knowledge to all corners of the earth, are taken one step closer to reality with HDTV.

In this shrinking world, impediments to the learning of basic medical information will not be acceptable, nor should they be. The essential truths that must be absorbed and understood do not recognize any such barriers. It is in the best interest of peoples of all nations to have the providers of their health care supplied with knowledge that is both uniform and accurate.

There is great worth to great pictures, and even one picture is said to be worth a thousand words. Then how much more is a picture worth that is five times better? Medicine, through its providers and its receivers, will surely be one of the major beneficiaries of this new technology.

# 26

# High-Definition Television and the Graphic Arts

*Murray Oles*
*Vice President,*
*Applied Graphics Technologies*

S ince 1975, when the graphic-arts prepress industry embraced digital technology for color separation scanning (that is, making color art and
photography into film suitable for use by a printer), the
industry has experienced one computer innovation after
another. Scitex, an Israeli-based textile company, was
the first to use digital color video technology to display
and manipulate a 32-bit (eight-bit per color—cyan,
magenta, yellow, and black) color image.

At the time, the only available frame-buffer technology was 512 by 512 pixels. That is, 512 by 512 pixels,
or dots, could be seen on the screen at one time. This
situation was a far cry from proposed HDTV standards.

Early system users thus experienced considerable
frustration in not being able to display on a video
monitor all the data included in the computer. In order
to maintain printing resolution requirements on the
computer output, the size of a full-screen computer-
based image when printed had to be slightly less than
two inches square. Larger images were subsampled,
creating a number of two-inch images to make up the
full-sized picture. A typical magazine advertisement
requires a total bit map of approximately 2,500 by 3,300
pixels; therefore, for every pixel displayed on the

prepress system screen in the early days, there could have been over 40 more in memory that were not displayed. In fact, the image on the screen had so little information compared to the image in the computer that even simple image placement on a page was a hit-or-miss proposition.

The manufacturers quickly realized that they would have to develop a technique that allowed the user to zoom in on one portion of the total image at a time. The user would thus be able to work on a small, but more accurate, portion of the image, tediously working from one tile, or image area, to another in order to achieve the quality of workmanship necessary.

With this system of virtual memory, however, each tile had to be saved separately before moving on to another tile. Therefore it was always necessary to carry out high-resolution image manipulation as a post-processing operation—when all the work was completed on individual tiles, the full-resolution image was created. Final page processing often took hours to complete.

In 1983, Crosfield Electronics, a UK-based company, developed its page composition system around a 1,024- by 1,024-pixel frame buffer. Crosfield believed that this higher screen resolution would be sufficient to avoid the necessity of developing a tiling system; in other words, the entire image could be manipulated at one time. This was true for images up to approximately six inches square. Beyond that size, customers began to have problems.

The Crosfield system brought many harsh realities to light concerning the manipulation of digital data within a higher-resolution frame buffer. The time it took to recall images from the computer's memory was considerably longer, the cost of the hardware was higher, and the monitors themselves were not really bright enough and were difficult to calibrate.

Today all the traditional prepress vendors operate on 1,024- by 1,024-pixel resolution screens. Scitex, Hell, Crosfield, and Dianippon Screen all offer a variety of page-composition and image-manipulation systems that employ virtually the same size frame buffers and operate in a similar fashion. All the systems have employed acceleration technology to speed up a variety of image-filtering functions to get the image on the screen, and all have developed a "write-back" technique where the edited portion of the image is written back over its original data upon completion. The write-back allows an operator to isolate a frame buffer-sized image area and work on it in real time, as opposed to the long processing time required with virtual-memory systems.

In order to speed up many of the operations on these systems, operators switch to a low-resolution display mode, which naturally results in a poorer quality image for the viewer. These same prepress vendors are currently developing systems that implement algorithms to display high-quality eight-bit color, which allows a fairly good image to be seen on the screen, yet speeds the system's operation up immensely. This type of display becomes less suitable, however, for delicate color manipulation work.

In contrast to traditional prepress vendors, Quantel Electronics has taken a much different approach to digital image manipulation with its Graphic Paintbox, which it developed after capturing the video industry with its Video Paintbox. The Graphic Paintbox employs HDTV technology to display the greatest amount of the image possible, and utilizes dynamic random access memory (DRAM) instead of traditional bit processing to build a very large frame store that is even larger than the display memory. This allows high-resolution even on close-up views of parts of the image.

Since Quantel developed its system later than traditional prepress systems, in 1986, it had more relevant technology at its disposal. The Paintbox has two frame-stores, allowing the operator to switch instantly between two images. This DRAM architecture, however, does limit the size of an image that can be brought onto the system—large-format print applications, depending on their subject matter, often require more data than the space available in a standard Graphic Paintbox.

To alleviate this problem, in 1989 Quantel released an XL version of the Graphic Paintbox. With approximately four times the DRAM than the original system employed, the 200 megabytes of memory now allowed an operator to work on a standard magazine spread at the same resolution as the printed piece will have. The XL version is capable of producing a commercially acceptable image up to even the largest billboard size. There is no post-processing involved after the work is done; therefore the Paintbox is a "what you see is what you get" (WYSIWYG) system.

The combination of Quantel's intuitive software interface and high-definition video display technology has provided the artist with a system that he or she can relate to. Studios that offer Quantel Graphic Paintbox also find that their clientele are drawn to the process of creating in real time.

With the Graphic Paintbox and its approach to meeting the needs of an artist, the shape of the graphic-arts industry is changing. Much of the traditional craft associated with making color separations has been embodied in technological solutions. The complicated nuances involved in properly stripping images on film to meet a variety of press characteristics are being automated. Electronic imaging for the graphic arts is placing the control in the hands of the originator of a design, instead of in the hands of the color separator as in the past. The digital data that is produced by the Paintbox can be accepted by any prepress system currently available.

The next generation of prepress system will contain more automation and be more like a Quantel Graphic Paintbox. Of particular importance to new development is the realization that the traditional scanning process need not capture a four-color signal, as it had done in the past. The real reason that prepress systems currently deal with four colors (cyan, magenta, yellow, and black) is that they were developed after color scanning—since printing in color required these four colors, traditional prepress systems followed suit.

Current color scanning technology converts a red, green, and blue (RGB) video signal to the cyan, magenta, and yellow (CMY) needed for printing. Simultaneously, grey balance and the black separation are added to the signal in real time.

Scanning technology became digital with the introduction of Crosfield's Magnascan Model 550 in 1976 and with the invention of digital image enlargement. Common sense dictates that input scanning and system manipulation should be done using the least amount of data possible to meet the quality requirements. A tricolor system can adequately describe all that is necessary to reproduce any image. Introduction of a fourth color, black, to facilitate the limitations of the printing process can easily be carried out during the exposure of color separation films, rather than during the original input scan, as in current prepress systems. Low-cost scanners and intelligent color separation film recorders are currently under development. Their release will open the door to video system developers that currently lack the knowledge necessary to develop a color electronic prepress system, thus allowing these developers to produce usable film for printing without going through a third party. The price of DRAM will also continue to tumble for the next several years, making it more economical to build large-capacity framestores and Paintbox-like systems. This will focus a much more critical eye on the display monitor.

The RGB phosphors currently used in video displays are not the best hue combination to create a good match with the printed image. It is probable that an industry-wide standard for RGB phosphors used on prepress systems will be developed in the next few years, so that the designer will be able to look at the video display and know that the color seen there is the same color that will be seen on film.

This calibration of the HDTV monitor to a printed sheet is not a trivial task. Because the red, green, and blue color phosphors are fixed, only the cyan, magenta, and yellow colors can be manipulated through software. Current methods of precisely evaluating transmitted monitor color are cumbersome. A method of developing a true three-dimensional look-up table, whereby color proofing could be done by the computer without a visual check of a printed press sheet, has not yet been perfected. These problems have been identified, however, and are being worked on. Once resolved, the monitor will be a far more acceptable medium for prepress proofing.

It is also worth noting that the printing process itself is undergoing a technical revolution as well. In 1989, Canon showed a plain-paper copier with a direct digital interface to the Quantel Graphic Paintbox. Developments are also underway that will utilize the benefits of liquid toner and electrophotography to develop a direct digital color photo copier capable of the highest quality at impressive speeds. For many short-run applications the designer will ultimately be able to obtain high-quality color printing through a local quick print service bureau. The transfer of digital data to the service bureau will be possible through modem transmission and high-speed communication services.

These developments are going to bring more and more demands on the performance of high-definition video display monitors. The prepress industry will want even higher resolution, and is already eyeing 2,056- by 2,056-pixel resolution displays for use in industry. In this case, however, more may not really be better. Rather than simply increasing the number of pixels, it is important to standardize on the next generation of HDTV display monitor so that developers can concentrate on perfecting color accuracy, which is not available in current HDTV technology.

Current paint system developers cannot compete effectively for the top of the market with anything less than HDTV. As an early Quantel user, and a pioneer of electronic prepress systems, I believe that HDTV has the resolution necessary for print applications. Future developers could serve us well by perfecting calibration procedures and methods of verifying color between monitors, rather than trying to further increase resolution.

It is interesting to note that as of summer, 1990, there were almost 30 Quantel Paintbox systems currently in operation throughout the United States. This loose confederation of independent companies represents the largest working HDTV network in the country.

# 27

## The Making of a Profitable
## HDTV Company

*Ronald Ratner*
*President,*
*Club Theatre Network*

*T*he first camera cranking away years ago heralding the birth of motion pictures was an unbelievable achievement. Imagine, images that moved—people around the world were amazed! Later, with the advent of sound, it seemed that there was nowhere else to go—all that could ever be accomplished was; the inventive forces had delivered the ultimate. But creative and inventive minds wouldn't stand still.

Next came motion pictures projected on higher and wider screens, glorious color, and improved sound techniques. And, along the way, gimmick ideas like 3D emerged that couldn't gain public acceptance because of severe technological problems. Good or bad, it was all part of the inventive process.

Somewhere in those fledgling years, actually in the middle 1940s, the words "television set" became a household word. There it was in black and white: moving pictures right in your home. Not film but images produced by a tube.

Next came color, larger sets and better reception, instant replay, and videotaped and live programming, brought into living rooms throughout the world.

Improvements in TV technology were extraordinary, but a breakthrough, something totally new for television, was needed. A clearer picture, a sharper image, colors that dazzled, but more importantly, television itself had to improve. HDTV, or as it's called in Japan, Hi-Vision, will be the answer. The world of television will never be the same: HDTV is the quantum leap the industry has been waiting for.

Along the way, the road to consumer acceptance worldwide will be bumpy. Lines will be drawn by government agencies in an effort to protect the consumer from laying out hundreds of dollars to make their television reception at home much better. This improved reception can and will be achieved with the higher-resolution picture of HDTV.

Already there are cries from consumer watch agencies that the public will be the scapegoat, and will be forced to pay higher prices to cover expenses for research, development, and marketing to take HDTV to the consumer marketplace. As we have seen in the past, on every technological front, whether it be appliances or broadcast tooling, initial consumer prices historically have maintained a high level, until public acceptance and demand have lowered their costs appreciably.

There are many other obstacles aside from the consumer question that must also be addressed. Basically what I am referring to is time and money. The use of HDTV production and editing in the studio and transmission capabilities will necessitate new equipment, hardware, and software, as well as the most important of all our resources, people. These are the people who will have been there from the beginning gaining the knowledge, and more will be needed who will have the ability to bring us this wonderful new medium.

As an example of pointing out the high costs: Today's one-inch tape (predominantly used in video production) lasts anywhere from one to two hours and sells for as low as $55 a reel. HDTV tape lasts only 63 minutes but costs about the same. At today's prices, we're talking about more than doubling the price of tape alone, if we are fortunate enough to get a good price. Tubes in today's standard broadcast camera last almost four times as long as today's HDTV camera tubes and cost approximately one-seventh the price. Yes, we are talking about significant costs all along the way.

Once the consumer fully understands the tremendous breakthrough in technology that has been made, however, not only in terms of better pictures but better sound, they will perhaps be prepared, socially and psychologically, to want HDTV. Remember, it really was just a few years back when color TV became the social necessity in neighborhoods around the world.

The HDTV industry is going to have the same effect as color TV did, no matter what negatives are perceived, because this is truly a better technology. It's a move forward that we believe the world will accept.

While the HDTV industry in the U.S. is scurrying to determine standards, analyzing costs of various systems, and competing for status in the standards battle, we at Club Theatre Network are already "turning on the lights."

The electronic theater and HDTV will together play an important role, indeed, in the emerging HDTV industry as well as serve as catalysts for the future development of HDTV and related industries.

For the record, I am not a television engineer, technician, or researcher in advanced electronics, but simply a businessman, an entrepreneur if you will, who believes that those of us who are fortunate enough to be in on the ground floor of this budding new industry have a tiger by the tail, and once the corporate and governmental battles end and the dust settles, the risks that we are taking today will seem minimal compared to the success that we will have realized.

With that in mind, the Club Theatre Network concept combines 35mm film quality, fiber-optic transmission, and computerized theaters with gourmet meals, entertainment, and live-audience participation and interaction. The results bring a new dimension of excitement to the traditional "evening out" of dinner and a movie, a live play, or a show. When CTN was originally formed as a viable company, our management team and board of directors had the common belief that the United States, because of its sociographics, was ready for a true breakthrough in the application of entertainment experiences. The original concept was to offer a fresh new idea: to provide an evening of entertainment in a very private and secluded atmosphere. Our initial programs included first-run motion pictures and gourmet dining in a high-tech mini-theater, all within the confines of a private country club or hotel.

At the outset, our research indicated that the audience demographics were the affluent, 40-plus age groups who could well afford the CTN concept of $35 and up per person for an evening of entertainment. These diners and theater-goers were presented with the option of dining and seeing first-run motion pictures in an environment that was not duplicated in any commercial theater. Being located in the private, safe, controlled environment in which the audience chose to live made the concept even more viable.

Originally, our method of technology for presenting this entertainment was through enhanced NTSC (an improved version of current television). Due to all the exciting advances in technology, it was possible for us to move into an area of sophisticated technology where no one has been before. Our prototype concept, introduced in June of 1985, included the projection of an enhanced NTSC signal transmitted by satellite. However, after my first introduction to HDTV technology, a whole new direction for CTN evolved. The prospect of high-definition television may immediately conjure up images of NHK, Sony, Hitachi, Toshiba, and other giant corporations. However, with the billions of dollars to be generated, smaller developmental companies will benefit from the work of these pioneers and emerge into the HDTV arena. The potential rewards for such enterprises are huge: big reputations, rapid sales growth, and a chance to make a mark in electronic business history.

While the CTN concept still involves dining and entertainment, how we present it has dramatically changed. It has become two-tiered to include production and post-production as well. Our plan is to have remote programming downloaded via satellite to the production studio and then delivered to the theaters via fiber-optic cable.

As of June, 1990, CTN is the only commercial

organization in the world to own the technological equipment (the Rank Cintel MKHD III High Definition Flying Spot Telecine) that makes possible the transfer of 35mm and 16mm films to the HDTV format. The telecine was delivered to our studio in Pompano Beach, Florida, at the end of February, 1990.

Each Club Theatre (as of June, 1990, there is one Club Theatre completed, with five scheduled to start construction in the summer of 1990) is a private electronic theater that receives a high-definition television signal from the CTN studio. This signal will be transmitted over our fiber-optic network, the establishment of which is, as of June, 1990, in the final stages of planning with Bell South, the regional telephone company. Once established, CTN will be the first chain of computerized HDTV theaters in the U.S. The theaters will be installed in exclusive country-club communities, high-rise condominiums, and luxury resort hotels, at first throughout the state of Florida. Expansion of the network is projected to include locations throughout the United States and eventually internationally.

The network will provide quality entertainment to wealthy and sophisticated audiences, capturing a segment of society, which, according to film-industry market studies, is generally considered to be among the least frequent movie-goers. As an exclusive guest at CTN, one can spend an evening or afternoon enjoying sumptuous cuisine, and after dining, be escorted to a private theater. There, the guest will lounge in plush, oversized seating, each seat equipped with its own microprocessor and handset, while enjoying a first-run feature film. Security, privacy, no lines, and no rushing are among the benefits of such a system.

Performances will not only encompass first-run motion pictures, but also live events, hit stage shows, ballets, plays, concerts, lectures, sporting events, and live performances with audience participation and computer-extended interactive capabilities. The computer-extended capability permits each member of the audience to actually participate in live performances and other entertainment experiences, for example, being the highest bidder during an international auction of fine art, jewelry, antiques, or even horses.

CTN was a host of an exclusive HDTV presentation of the Sugar Ray Leonard–Roberto Duran fight at the Gusman Theatre of the Performing Arts in Miami in December, 1989. This venture was held in conjunction with NHK Enterprises USA (providers of the HDTV transmission), Southern Bell (fiber optic provider), and Barco (provider of a projection system that has the capability to project an HDTV picture of this size). This was the procedure: NHK Enterprises USA broadcast a live signal from Las Vegas and transmitted it to a Hughes communications satellite. The HDTV signal was then down-linked in Miami and decoded at the Southern Bell switching station. The signal was then placed through a MUSE decoder and sent approximately three miles over a fiber-optic cable to the Gusman Theatre. There, we projected it through the Barco projection system and produced a picture 29 feet wide by 17 feet high.

Those attending said they were amazed at the picture clarity and the quality of the sound system. Most seemed to feel that it was the closest thing to actually being at the fight. This was the first commercial transmission and use of HDTV in the U.S. We certainly hope to be doing more of these large-scale events.

My personal feelings are that in the not too distant future, all of the major theater chains will be utilizing the HDTV electronic technology. The use of HDTV and fiber-optic delivery will give Club Theatres and other HDTV theaters broader uses as well. For example, shareholders of a company would be able to attend an annual meeting without having to travel to the city in which the meeting is being held. Instead, they would sit in the Club Theatre nearest them. The handsets would allow the shareholders to ask questions and the keypad to cast votes. Buyers for clothing companies could attend a video merchandise mart, see the latest fashions, and record purchases with the keypad and handset. Fashion shows could be brought to the country clubs, condominiums, and hotels. Via satellite, they could come from anywhere in the world and allow people to make their purchases through the Club Theatre interaction seat. We also have plans to do several live performances including an ice show.

The admission price for the dining and entertainment package is expected to range from $35 to $100 or more per person, depending on the featured programming. All additional items purchased during the performance can be charged directly to the patrons' major credit cards. Between receipts from the dining and entertainment package and the revenues generated from our various marketing applications, our program hopes to prove economically feasible to all of our advertisers, merchandisers, franchises, and the network itself.

Although CTN intends to own and operate at least 14 theaters, many other theaters are expected to be franchised or subleased. Franchising is expected to provide a substantial amount of revenue and is a major part of our belief in the financial viability of the CTN concept.

CTN also has support operations, which were developed to supplement the HDTV network and provide additional profit centers: CTN has purchased an established NTSC production facility located in Pompano Beach, Florida, called Media Productions, as well as the 28,000-square-foot building it is housed in. Together, CTN and Media Productions will form an integrated HDTV company. The new company will handle production, post-production, film-to-tape transfer, and serve as the home base for transmission of the HDTV signal throughout the network.

The purchase of the HDTV Rank telecine from Rank Cintel of London, England, was of paramount importance to the network in order to provide continuous product to our theaters at a reasonable costs. Not only will the Flying Spot telecine be used to transfer first-run movies for our own HDTV theaters, but it will also be available for transfers for outside production companies as well, especially in the area of special effects. CTN is the first company to use this machine commercially. It insures the success of the image quality for our HDTV presentations.

Initial contacts have been made by the HDTV studio with companies including Turner Entertainment, MGM/UA, and Southern Bell, to discuss providing production-related services to these companies. Additional referrals are being handled on a fairly steady basis. Southern Bell has surveyed approximately 25 locations that can eventually be connected to our studio in Pompano by fiber-optic cable.

There have been tremendous strides in HD technology made by the Japanese and European video equipment manufacturers, much of which has been seen at various trade shows. Yet all of the emphasis has been on hardware and very little attention is paid to software. What good is all this wonderful equipment if all you can watch on it are clips of *Top Gun* and *Back To The Future* at a trade show?

With the rapid growth of HDTV both in the U.S. and abroad, the demand for software is expected to be enormous. We see a large part of our position as a major provider of that programming. CTN intends to work together with the film industry to further both the film and HDTV mediums. We do not agree with the theory that HDTV will eventually replace 35mm film. Hollywood, after all, will continue to be a major provider of both inventory and new product for both formats.

NHK is now transmitting one hour per day of HDTV at selected public sites throughout Japan and has intentions of increasing that to several more hours in the near future. Once they begin to transmit several hours of HDTV per day, the existing supply of programming will not last long. Accordingly, libraries will be a valuable commodity. In the United States, we liken Club Theatres to NHK's campaign; after all, CTN will be the first place American consumers will get their initial taste of HDTV, and our target market will certainly be able to afford the new equipment. After viewing HDTV at our theaters, the next step will be for them to purchase HDTV VCRs and HD monitors and projection units.

By teaming with companies like NHK, Sony, Southern Bell, Rank Cintel, and Barco, whose philosophy is to further technological advancement in our industry, we have assured a solid foundation for Club Theatre Network. Solid working relationships with firms like these will help us realize CTN's short-term plans—14 theaters operating in South Florida within the next 12 months.

Earlier I mentioned the very beginnings of motion picture and broadcast entertainment. It is true now, as it was then, that technology, research, creative thinking, and taking chances are all part of burgeoning concepts created and developed by men and women who believe in ideologies and dreams.

At Club Theatre Network, we are an extension of these beliefs and challenges. We're excited that in some small way we are going to contribute to the technological advancement of HDTV, which we believe is now on the threshold of greatness.

# 28

## A Few Reflections on
## High-Definition Video and Art

*Barbara London*
*Video Curator,*
*The Museum of Modern Art*

*T*his is an exhilarating moment on the eve of the
21st century. The versatile world of electronics is changing more radically than was ever
imagined, with smaller and faster computers poised to
become a more integral aspect of public and private life
than they are now. What have been seen as separate
devices—TV sets, tape recorders, disc players, speakers,
computers, and fax-cum-color copiers—are more compact and powerful than before, and capable of operating
smoothly as integrated systems. Today it is the users
who design and determine applications for their "multimedia" hardware. Accessing information quickly from
their own archives as well as from outside databanks,
they can process, file, send, exhibit, or publish to their
own specifications, meanwhile watching and/or listening to other events as inserts on the same screen or
soundtrack, and answering electronic mail. Central to
these flexibly expanding operations is the need for a less
cumbersome video display system (ideally a flat screen)
with sharper, digitally controlled images and film-qual-
ity projection.

In the United States, at least, it will be a while before
high-definition hardware actually reaches consumers,
given the sensitive national policy questions surround-

ing manufacturing and the establishment of technical standards. From past experience we know that politics more than market demand determine how, when, and which electronic systems take hold. In the United States, which has shied away from a clear telecommunications policy, where tools are studied more easily than goals, high definition is currently being used as a new production tool. The software must then be converted to NTSC (current) video and/or 35mm film for more widespread use. With so much technical change, worlds that until recently have been very separate are slowly merging, forming new financial, technical, and creative alliances. These spheres include the digital realm of computer graphics; the influential world of producers and publishers; and the distribution channels, including broadcast, cable, satellite, and home videocassette and disc.

Really a new medium, in look and application high definition falls somewhere between film and video, with its seemingly "live," detailed image, which has to be "played" to be viewed, its capacity of "instant replay," and its multiple collage-like layering of material. Embodying the best of the "time-based" media, high definition allows distributors to send high-resolution sound and video- or film-derived images electronically through the same system and enables graphics more easily to be used interchangeably between still and motion applications. It also allows manufacturers to sell more equipment, as this new technology requires a substantial hardware investment, either as a new installation of gear or as replacements of older technologies. High definition will touch many aspects of contemporary culture, especially the publishing and entertainment industries, as well as art making and education. In time, high definition will have an impact on the creative aspects of image making. New technologies (like scientific discoveries) generate the excitement and energy that stimulate artistic talent to produce breakthrough solutions. Because production equipment is currently all but in the prototype stage and is therefore very expensive, the only people with access to the hardware at present have at least a tangential connection to the commercial world. A lot of the initial work in high definition can be characterized as thoughtful technical problem solving. Only when more artists have had adequate hands-on experience with the tools and have explored the multimedia applications will we begin to discover what is truly unique to the medium. Only then can we go beyond Hollywood and commercial television formulas.

At the moment there are a handful of programs that are breaking new ground and leading the way. An active proponent is the Polish-born animator Zbigniew Rybczynski, who in 1985 became attracted to high definition for its clear, film-like image and its capacity for extensive layering. A virtuosic craftsman, Rybczynski carefully structures his projects, accurately repeating his camera movements during shooting so that in the editing room he can duplicate, manipulate, and then superimpose his performers' actions over separately recorded background footage. His creative process is most closely analogous to that of the sound recording artist. Rybczynski's most recent production, *The Orchestra* (1990), was coproduced by "Great Performances" of New York's station WNET/Thirteen; French station Canal Plus; and Japan's national network. Such international coproduction alliances are necessary today, given the exorbitant costs of cultural programming and the fact that high-definition hardware is still extremely expensive as it moves beyond the prototype stages.

Inspired by six familiar classics, including Mozart's "Piano Concerto No.21" and Ravel's "Bolero," Rybczynski cleverly scored his *Orchestra* visuals to symbolize political repression and freedom. He superimposed his performers' actions over the ornate Paris Opera House; Chartres' soaring cathedral; Delacroix's 1830 painting *Liberty Leading the People;* and an endless staircase set against a dramatic sky, reminiscent of Eisenstein's film *Potemkin* (1925). Rybczynski draws upon music's subliminal force to examine mischievously such universal themes as sex, social mores, and power.

Another creative advocate is Hideaki Maekawa, an innovative and prescient producer/vice president at Tokyo Broadcasting System (TBS) in the Advanced Systems Department. Working with a talented team that includes experimental film maker Kohe Ando, Mr. Maekawa has created two award-winning high-definition productions. The first, *Mahoroba* (1988), recounts the medieval "Legend of the River Ikuta," about a noble young woman forced to choose between two equally matched suitors. She tells them that the man whose arrow first hits the swan floating in the river will win her hand. When both simultaneously strike the same bird, she takes her own life, followed by the two men. For their setting, the producers chose a remote snowy landscape in the Oku-Nikko area. Rather than push the wide range of color that can be achieved in high definition, they concentrated on the subtleties of white. They found the extremely bright and reflective outdoor light to be reminiscent of the subtle contrasts between black and white found in traditional Japanese sumie ink drawing. Their second program, *My Small Table* (1988), goes in an entirely different direction. It is an ironic celebration of food, a takeoff on the ubiquitous television cooking show. Based upon a Tokyo artist's diary recounting his sophisticated exploits with food, the producers set the sensual, darkly textured program in his westernized home, where the luxurious tableaux of food being prepared and relished resemble 16th century European Mannerist paintings. These wry explorations in high definition celebrate life as well as being clever technical exercises.

The Hi-Vision Department of Sony Japan produced *Infinite Escher* (1989), an eight-minute surreal story about an alienated adolescent befriended by a human-headed magical bird. They then together enter an imaginary architectonic place. The project blends the geometric fantasy world of M.C. Escher with everyday reality by combining state-of-the-art 3D computer graphics with high-definition footage shot both on Manhattan's streets and in the studio. An ingenious pictorial blending of artificial and actual imagery, the program was made with all new equipment: superminicomputers, the latest HD cameras and decks, readable-writable-erasable optical disks, and specially designed frame buffers and software. *Infinite Escher* was written, designed, and constructed by a talented creative team in

New York, including John Sanborn, Mary Perillo, Dean Winkler, and Barry Rebo. For many creative people working in high definition and computer graphics, the goal is to construct even more believable fusions than have yet been produced, and more realistic artificial renderings. This is becoming possible with digital hardware.

For our cultural institutions today, survival also means adapting to our electronic age, using computers to manage vast amounts of archival information, to produce didactic materials quickly, and to operate more efficiently as businesses. These complex, multifaceted organizations act as archivists committed to preserving art traditions; as showcases for new forms of artistic expression; as teachers of art appreciation; as record keepers; as databases; and as publishers of books, slide sets, films, tapes, and discs. Wanting to expand their audiences (and, in the United States, to reflect the country's broad cultural diversity), these cultural entities are realizing that multimedia is the most expedient route to take. But change, which is a political, economic, and time-consuming operation, begins slowly, especially in the nonprofit sector. Operating with limited cash reserves, cultural organizations must utilize their valuable archive-cum-database through thoughtful multimedia and multicultural programs. New alliances with corporate entities have to be explored; both the cultural institution and the corporation need to develop prototypes that will lead to more effective educational programming, greater outreach, new income, and more visibility. Potential applications of new technologies, including the denser image of high-definition video, need to be understood. This takes experimentation. Museums will move slowly. It is a world where traditional print reproduction remains strongly entrenched, where image quality is closely scrutinized for its faithfulness to an original.

An excellent example of this exploration is the Museum Education Consortium, a collaborative project between the education departments of seven museums: The Art Institute of Chicago; The Brooklyn Museum; The Metropolitan Museum of Art, New York; The Museum of

Fine Arts, Boston; The Museum of Modern Art, New York; The National Gallery of Art, Washington; and The Philadelphia Museum of Art. Funded by the Pew Charitable Trusts, the Getty Grant Program, and the Warhol Foundation, and overseen by a talented design team headed by Susan Stedman and Kathy Wilson, the Consortium has created a prototype multimedia teaching system for two principal constituencies. The first constituency includes adult museum visitors, who are curious and interested, but are often uncomfortable with and unskilled at looking at art. The second constituency is comprised of professional teachers, who in turn address audiences at all levels of schooling, and therefore all ages. The prototype consists of an extensive databank containing images of Impressionist and Post-Impressionist works of art from the seven Consortium museums, as well as other significant collections; many documentary images related to the same period; moving-image and audio source material, such as music and narration; and, finally, relevant textual information (timelines, maps, bibliography, etc.). This image and databank is the visual resource upon which the two computer programs (one for each constituency) draw. The development of new imaging systems, software, and other aspects of technology will lead to new audiences, unable to be identified or estimated for size at present.

For economic reasons, and because of a mandate to come up with a well-designed pilot that could be available in 1990, the Consortium had to design the project for NTSC (current television) disc, for interactive use with a personal computer. However, as a part of the pilot, Cézanne's painting *The Bather* (1885) and Monet's *Water Lilies* (c.1920) from The Museum of Modern Art were taped in high definition by REBO Studios to explore a higher-quality image, which was then down-converted to NTSC.

Today, our cultural institutions have the challenging opportunity of creating responsible multimedia programs that give intellectual access to new audiences. Electronic media in general, and more specifically high definition, are powerful tools that can improve understanding and aid the interaction of different cultures. They can and will bring our worlds closer together, in terms of visual as well as communication media, and can also help us address a myriad of issues that are part of cultural cross-fertilization. With more rigorous cultural exchange will come heightened awareness and responsibility for our shared future. To live means to change, and we are all capable of extraordinary growth.

HANDICAPPING THE FUTURE

# 29

## Paradise Lost or Found

*John F. Rice*

*I*n the next few years, decisions will be made that will affect television for the coming generations. The impact of those decisions could have broad implications on everything from America's balance of trade, to employment, to long-term industrial development and strength in areas philosophically and geographically far away from television and HDTV.

At one time, it was hoped that HDTV could be a worldwide television system, unencumbered by political, technical, or geographic boundaries. Today, with companies lobbying for their proposals before both technical and political committees with equal vigor, the one-world vision has been lost. It is probably true that issues of politics and technology would have made that goal of a world standard impossible under any terms. But it was once part of the dream.

What remains of the dream are two ideals: a technically superior system and an industry that creates maximum benefit to all parties playing the game. But both of those goals have already been compromised.

HDTV, when it is approved (in the U.S. for transmission, and in the world market for technical standards), will not be the absolute best system available from a technical point of view. It will be a compromise. The technical developments of the day must be accepted at some point, and we must move on or run the risk of constantly waiting for a better system. Technicians and inventors will continually create new devices and theories. They can always improve on the past. Already, there are computer systems that operate at greater bandwidths or higher resolutions than those proposed for HDTV.

Of the HDTV proposals currently on the table, both in the U.S. and throughout the world, all are vast improvements over current television systems. Like it or not, we are stuck in that realm of improvement. We will also be "stuck" with 1990s HDTV long into the 21st century. Whatever improvements, evolutions, or inventions are developed in the HDTV era, they will be built on the technology that is selected today. We need to take the best of what we've got before us now and bring it to the marketplace.

The political compromise may not be as easy as the technological one. Too many people, companies, and governments have invested too much time and money on systems that are fundamentally different from one another. In the final resolution, there will be winners and losers.

To the winners go the marketplace and with that, the profits and financial power of the future. The losers will be left with debts and plenty of time to figure out what they did wrong, if anything. Among the winners will be individuals, companies, and even countries. The stakes of the game are high. Not only will HDTV mean new television sets, but new television stations, new production equipment, new marketing channels (for consumer and professional gear), new manufacturing opportunities, and new import/export relationships.

Early efforts to develop worldwide agreement on HDTV formats and systems were based on a purely technical debate about how to create an ideal technology that fit everybody's technological needs. But as the world engineering community failed to come to a consensus, HDTV also fell victim to political and economic fights.

Today, the technical discussions are almost secondary to arguments over what HDTV means to national economic strength, world trade leadership, and political importance. When the discussions in the U.S. turned to Washington, DC, and foreign governments started funding their countries' development efforts, the technology itself almost became moot. In some quarters, like Congress, it appears to be more important to have an HDTV industry that creates jobs and makes America strong than to have the best possible technology.

Should the world have agreed to a Japanese-developed standard for the sake of speed and technology, or are individual governments and related partnerships correct in fighting for their nationalistic benefits, despite technology? That is a a question for the historians. The fact is that before there is decision on HDTV (or a series of determinations throughout the world), there must be compromise.

That any individual country has placed an emphasis on HDTV as a major factor in its economic future is both frightening and exhilarating. It's frightening because HDTV is the ultimate gamble. Nowhere in all the debates and discussions has there been any definitive promise that HDTV will succeed in the marketplace. There is no guarantee that the company or country who owns the technology will control the manufacturing. And there really is no way to defend against something better coming along and wiping out all that has been created to date.

It is exhilarating in that so much energy has been dedicated to HDTV. Has any technology been given such a public forum while it is still in its embryonic stage? HDTV is now a public debate, although an oft-confusing one.

There are too many elements of the debate that, at their cores, are diametrically opposed. Yet, the drive for the ultimate technology knows no political boundaries. The demand for national, or even individual, corporate success and the investments made on that behalf will drive a country or company to promote its own system. These political and economic proponents may be more driven by finance or emotion than by excellence.

If European nations are proposing their systems

because they are truly best technologically for the countries, that is one thing. If they are creating systems of their own so as not to be dependent on Japan or the U.S., that is another. Both arguments have elements of validity. Both are defensible.

Different countries have different technical demands, so there is reason for differences in television systems or, in fact, in any electronic system. And regardless how one feels about the goal of technological excellence, can you argue against any country that wants to put itself in the strongest economic position?

If there is a realistic goal for HDTV, it is to create a system or systems that best fulfill all of the individual demands of technology, economics, and politics. If there is a compromise to be made, is it possible to make it so it allows the maximum number of winners and minimum of losers? It is time for negotiation, not borne of altruism, but of practicality.

There is no requirement for worldwide consensus. It is possible that varying geographic regions could create and mandate their own independent HDTV systems, or even create multiple systems that overlap in their industrial and geographic domains. But enough people are still working toward an agreeable worldwide decision on HDTV for us to hope that their efforts succeed.

The world is waiting for HDTV. But the final decision on a system or systems is only the beginning of HDTV's realization. There are many variables to the ultimate success of HDTV that can't be played out until there is a sense of direction. Too many questions remain unanswered.

Too many people are waiting for government and standardization decisions before beginning their own decision process.

HDTV as a production medium for television, motion pictures, and other entertainment and informational media is a market that can really begin to develop only when there is a definition of HDTV. The current efforts to create markets for HDTV production equipment are as much political moves as they are real marketing efforts. To a large degree, manufacturers are trying to create current equipment markets as part of their argument for their own systems. These companies have already invested incredible amounts of time and money in developing these products and they certainly would like to start seeing some sales revenues.

But the majority of the production community remains unconvinced of HDTV's potential. Producers see no need to think about changing their minds while the overall standardization and transmission issues are unresolved. Why should someone produce in HDTV if there is no outlet for the product? Some companies are taking the gamble that their productions will be more valuable in the future because they already exist on HDTV; their choice may prove to be wise. Others have created productions in HDTV because the existing technology offers benefits to the production either in cost savings or in better quality, or because HDTV can do things that current video or film production techniques can't accomplish. But both the gamblers and the technical users are in the minority. For HDTV to succeed at its fullest, it needs to be accepted by the global production community. The production community won't move *en masse*, for any reason, until there is a transmission system.

The largest variable for the future of HDTV is the unanswered question of whether or not the consumer really wants it. There have been a smattering of studies that prove or disprove the arguments of a massive HDTV business. Will HDTV television set sales match the pace of current color TVs? Or PCs? Or CDs? Most likely the sales curve of HDTV will be unlike any other product. We don't know how fast that curve will rise, or if it will rise at all. We won't find out until there are products for sale.

HDTV has had incredible advance publicity. The concept that the HDTV industry will be strong enough and broad enough to impact any country's economy is both part of the publicity and caused by the publicity. Certainly, Japan sees the potential. Europe sees the pot of gold. And the U.S. also looks to HDTV as a solution to a lot of its problems.

It is interesting and disturbing that so many people

are investing their futures in HDTV, especially when there are so few guarantees. Can HDTV strengthen an economy, or save one? I don't know. But I can see huge gaps in many of the American arguments.

When the U.S. looks to HDTV to correct errors of the past in the consumer-electronics field, and envisions a future where HDTV transmission is a U.S. system, HDTV sets are made by U.S. companies, and the exploding semiconductor market is the domain of the U.S., the theories are wonderful. But there is no way of ensuring that the dollars from HDTV remain in the U.S. Why should any company, foreign or U.S., manufacture its products in the United States if they can produce them more cheaply overseas? Where is the law that restricts foreign-owned companies from making products that comply with U.S. standards? Where is the insurance that HDTV, U.S.-based or foreign-based, means great things for the U.S. economy?

Let's face it, HDTV could be the biggest boondoggle of the '90s.

Personally, I hope it is not. Although I think concerns over marketability and economic value are valid, I hope the negative assessments are wrong. I like HDTV.

Whatever the analyses, the studies, or the position statements say, HDTV is on the way. The decision will be made in a matter of time, hopefully faster than slower. The market reaction will remain unknown until then.

HDTV today is caught in the currents of its ambitions and goals. At its fullest, it can be most if not all of the things it has been predicted to be. Regardless, the best that can be done now is to make it a reality, and then watch what happens. There has been too much investment, both financial and emotional, not to play this out to its conclusion.